THE SECRET WORLD OF THE
BRAIN

METRO BOOKS
New York

An Imprint of Sterling Publishing Co., Inc.
1166 Avenue of the Americas
New York, NY 10036

METRO BOOKS and the distinctive Metro Books logo are registered trademarks of Sterling Publishing Co., Inc.

Design © 2016 by Carlton Books Limited
Text © 2016 by Carlton Books Limited

All rights reserved. No part of this publication may be reproduced, stored in a retrieval system, or transmitted in any form or by any means (including electronic, mechanical, photocopying, recording, or otherwise) without prior written permission from the publisher.

ISBN 978-1-4351-6272-3

For information about custom editions, special sales, and premium and corporate purchases, please contact Sterling Special Sales at 800-805-5489 or specialsales@sterlingpublishing.com.

Manufactured in China

2 4 6 8 10 9 7 5 3 1

www.sterlingpublishing.com

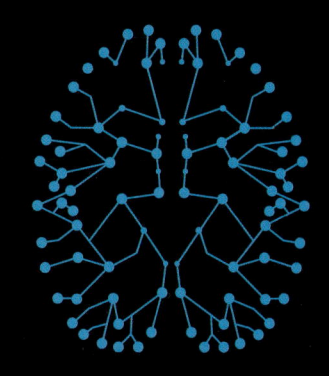

THE SECRET WORLD OF THE BRAIN

WHAT IT DOES, HOW IT WORKS, AND HOW IT AFFECTS BEHAVIOR

CATHERINE LOVEDAY

METRO BOOKS
New York

CONTENTS

7-9
INTRODUCTION: YOUR SPECIAL BRAIN

10-23
CHAPTER 1: UNDERSTANDING AND INVESTIGATING THE BRAIN

24-35
CHAPTER 2: HOW DOES THE BRAIN WORK?

76-87
CHAPTER 6: ARE FEMALE BRAINS DIFFERENT FROM MALE BRAINS?

88-97
CHAPTER 7: WHY DO WE CRY AT FILMS AND LAUGH AT JOKES?

98-107
CHAPTER 8: HOW DOES THE BRAIN TELL THE TIME?

142-153
CHAPTER 12: HOW THE BRAIN SPEAKS

154-165
CHAPTER 13: MAKING SENSE OF OUR WORLD

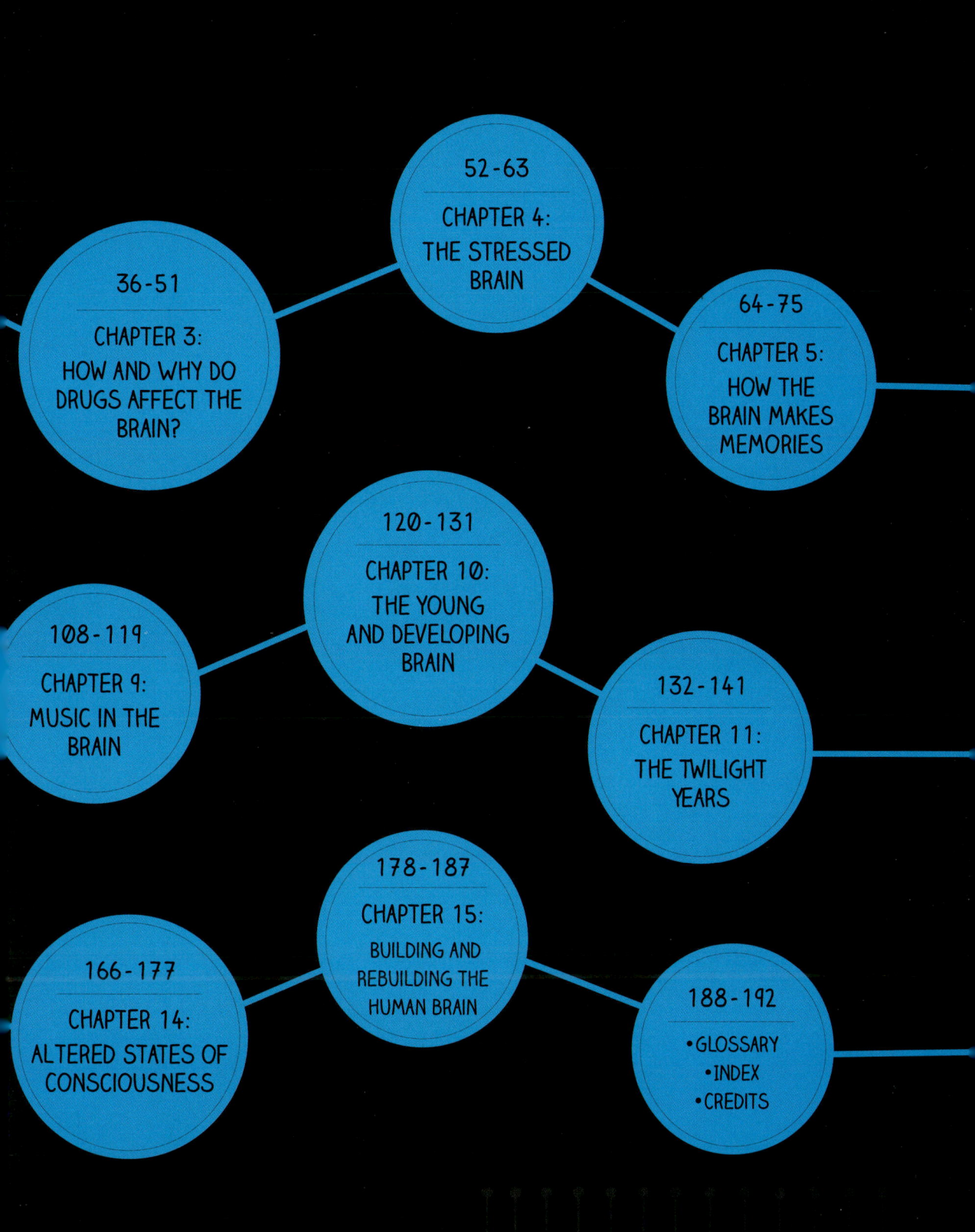

36-51
CHAPTER 3:
HOW AND WHY DO DRUGS AFFECT THE BRAIN?

52-63
CHAPTER 4:
THE STRESSED BRAIN

64-75
CHAPTER 5:
HOW THE BRAIN MAKES MEMORIES

108-119
CHAPTER 9:
MUSIC IN THE BRAIN

120-131
CHAPTER 10:
THE YOUNG AND DEVELOPING BRAIN

132-141
CHAPTER 11:
THE TWILIGHT YEARS

166-177
CHAPTER 14:
ALTERED STATES OF CONSCIOUSNESS

178-187
CHAPTER 15:
BUILDING AND REBUILDING THE HUMAN BRAIN

188-192
- GLOSSARY
- INDEX
- CREDITS

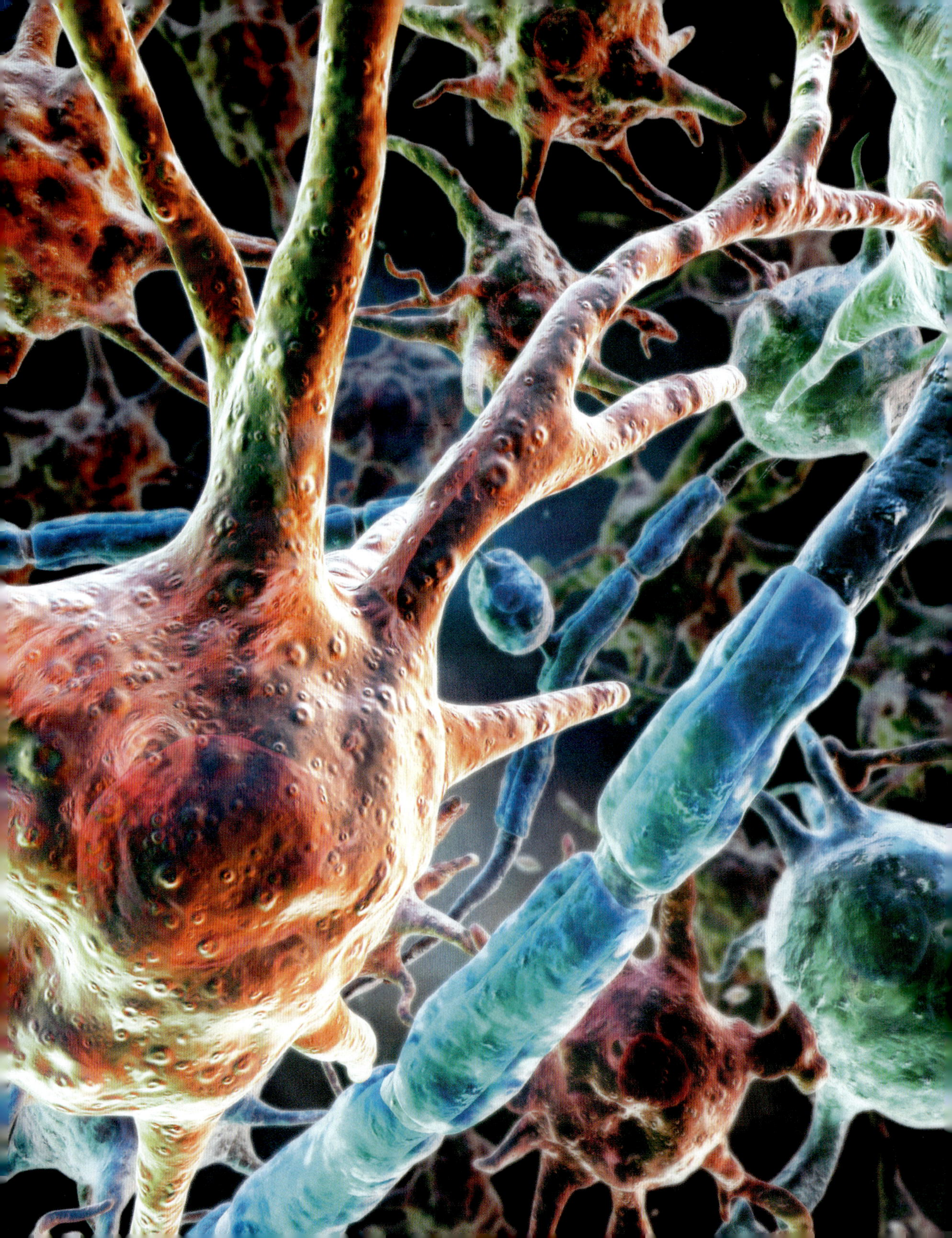

INTRODUCTION: YOUR SPECIAL BRAIN

HAVE YOU EVER WONDERED WHAT IS HAPPENING WHEN YOU SPONTANEOUSLY LAUGH AT A JOKE OR SUDDENLY FIND YOURSELF CRYING DURING THE HAPPIEST PART OF A FILM?

Do you ponder over why the smell of cut grass instantly transports you to another time and place, or why music can have such a profound effect on how you feel? Do you want to know why drinking coffee can suddenly wake you up, or why drinking alcohol can affect the way people behave? The secret lies in the workings of your brain….

COMPLEX GREY MATTER

The brain is the most complex organ in the human body and, as well as being the main control centre for most of our bodily functions, it gives us the ability to read a book, carry out calculations, recite a poem, remember our past, plan our future, taste a piece of chocolate, solve problems, experience joy and misery, fall in love, wonder at a piece of art, sing a song and countless other things that we take for granted every day. This 1.5-kg (3.3-lb) fatty lump of cells – which in its living state looks and feels quite similar to a rice pudding – is extraordinary. It is at the very centre of who we are, what we feel and how we behave.

Opposite: The human brain is one of the most complex structures in the known universe.

Below: The common earthworm has a version of a brain – a nerve centre that controls its movement.

Of course, human beings are not the only animals that have brains. Our favourite pets – dogs, cats, horses – all have well-developed brains and are capable of complex behaviours. In fact, even the common earthworm has an area in which all nerve pathways are coordinated, so we could describe this as a very primitive brain – a prototype, if you like. Compared with vertebrate animals, the worm has a very simple nervous system and so requires little in the way of higher-level control. The worm's nerves work together in much the same way that members of a small team may coordinate with each other on relatively simple tasks. In contrast, human beings have evolved to such a degree that they have an incredibly complex nervous system and, just as a huge corporation has a boss to manage all the workers and oversee all the necessary tasks, so the human body has a brain.

REFLEX VERSUS MANAGEMENT

To continue with this analogy, it makes sense for some tasks – those that are urgent and well practised – to be carried out very quickly with little or no input from the boss. Much the same is true for the human body: the automatic response that occurs when someone steps on a pin or touches something very hot is a "reflex" action. A reflex such as this consists of very simple connections located in the spinal cord and has no input from the brain. Most of what we do, however, requires extensive management of many different actions, as well as the ability to refer to past knowledge and events and imagine future ones. Put simply, the bigger an animal, the more complex its nervous system. By the same token, the more sophisticated its social skills, the more significant the role of the brain.

So the human brain is complex – but is it clever enough to understand itself? Philosophers have debated this paradox for centuries, with some claiming that a system can only ever understand another system that is less complex than itself. They argue that human beings will never fully understand their own brains and that to do so is like trying to smell your own nose. As one pharmaceutical worker is memorably quoted as saying, "If the human brain were simple enough for us to understand, we would be too simple to understand it."

THE BRAIN PROJECT

Nevertheless, in April 2013 US president Barack Obama announced the launch of the ambitious BRAIN Initiative, with an initial investment of $100 million. Inspired by the Human Genome Project – an international scientific research project concerned with the make-up of human DNA – the BRAIN Initiative endeavours to pull together many of the biggest worldwide research institutions in a coordinated effort to understand and map the human brain. The project is expected to last for more than ten years, require extensive development of technology and cost hundreds of billions of dollars. It has been greeted with scepticism by some quite prominent neuroscientists, who believe that it will divert funding away from important individual projects and that there is too much focus on the development of technology. However, many others feel that a united and concerted approach such as this has huge potential and may be the only way to combat many of the distressing and debilitating conditions that affect the human brain.

President Obama's project may or may not lead to a huge leap forward in our understanding of the human brain, but let us not underestimate how much we already know. The combined efforts and dedication of many brilliant thinkers – scientists, philosophers and doctors across the centuries – have given us an enormous amount of insight into the anatomy and chemistry of the brain: how it develops and ages; how it regulates our stress levels and what happens when that goes wrong; how we lay down and retrieve memories; what happens when we sleep; how speech is produced and understood; why music is so important to us; and why laughter makes us feel good. We may never fully understand the human brain, but we know a great deal more than we did one hundred, or even ten, years ago. This book will hopefully provide answers to many of your questions about the brain, but for every one it answers it will most likely provoke ten more. Such is the curiosity and hunger of the brilliant and fascinating human brain.

Opposite: So much is still unknown about the brain, but this incredibly complex organ is the focus of dedicated world-class research.

INTRODUCTION: YOUR SPECIAL BRAIN   9

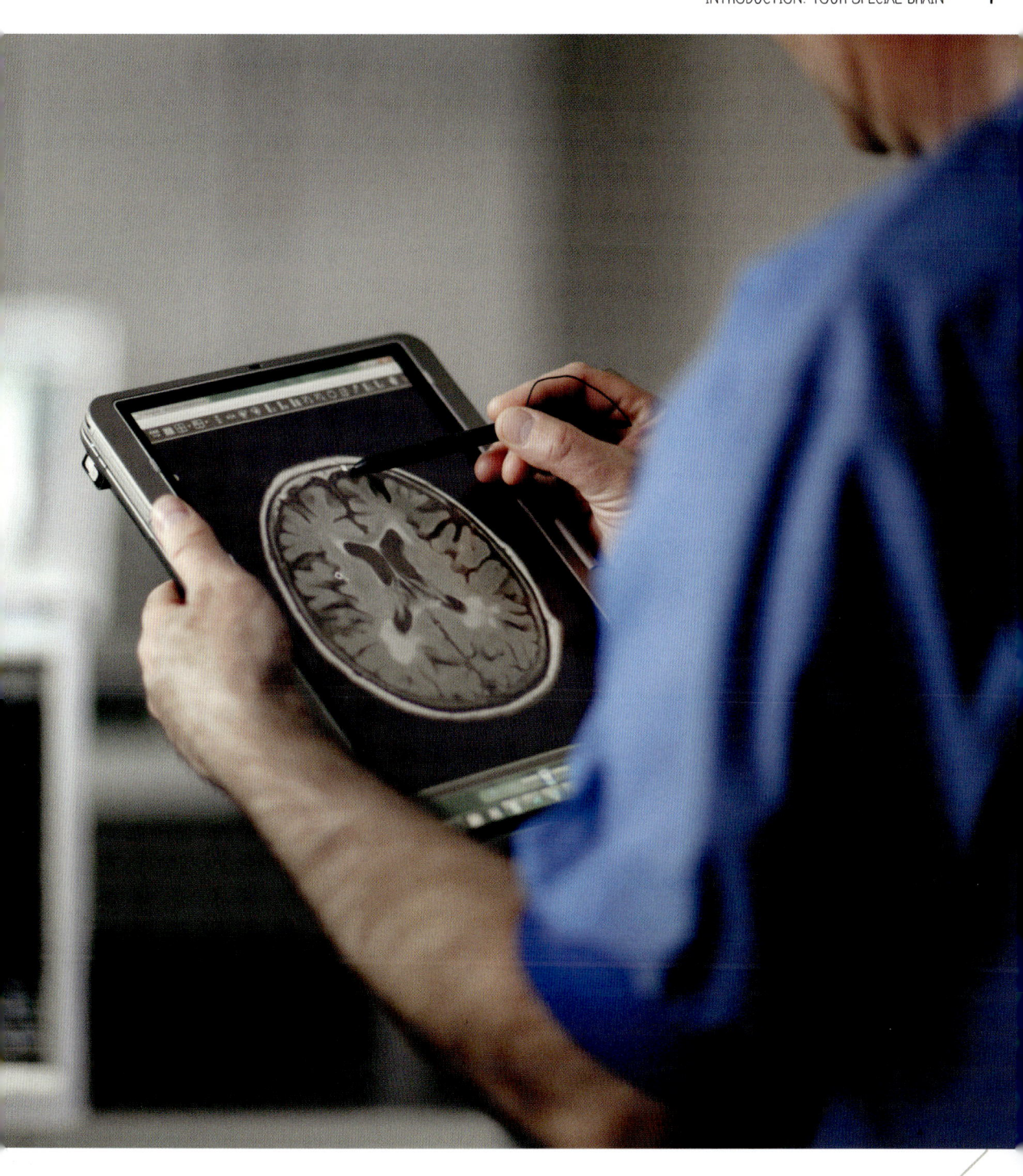

1. UNDERSTANDING AND INVESTIGATING THE BRAIN

THE GREAT ITALIAN RENAISSANCE ARTIST, SCIENTIST AND INVENTOR LEONARDO DA VINCI ONCE FAMOUSLY SAID THAT TEARS COME FROM THE HEART AND NOT FROM THE BRAIN.

This may be a beautifully romantic statement, but da Vinci was of course wrong: the brain ultimately drives all our behaviour. Tears and laughter perfectly illustrate the connection between our mind and body – they are a very physical manifestation of a psychological state. So, before we look in detail at the structure and function of the brain, it might be useful to look at how it all fits together and examine how the nervous system as a whole is organized.

Below: The human brain is one of the most complex structures in the known universe.

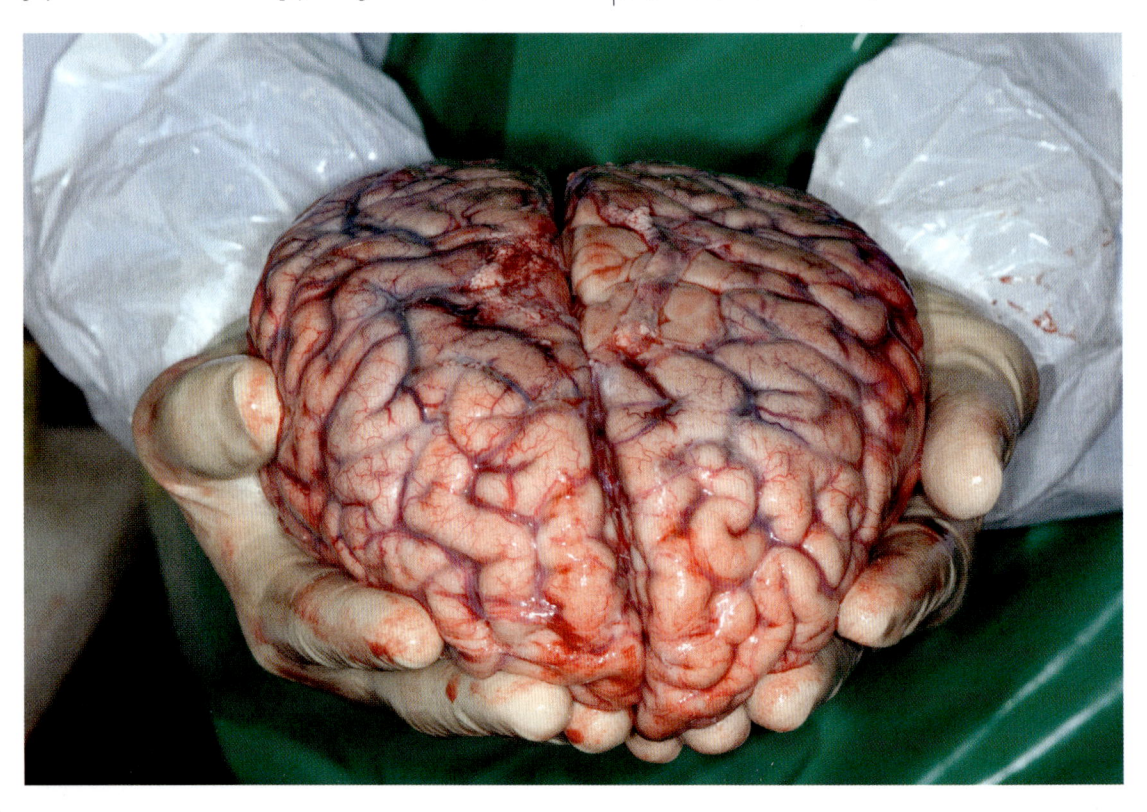

UNDERSTANDING AND INVESTIGATING THE BRAIN 11

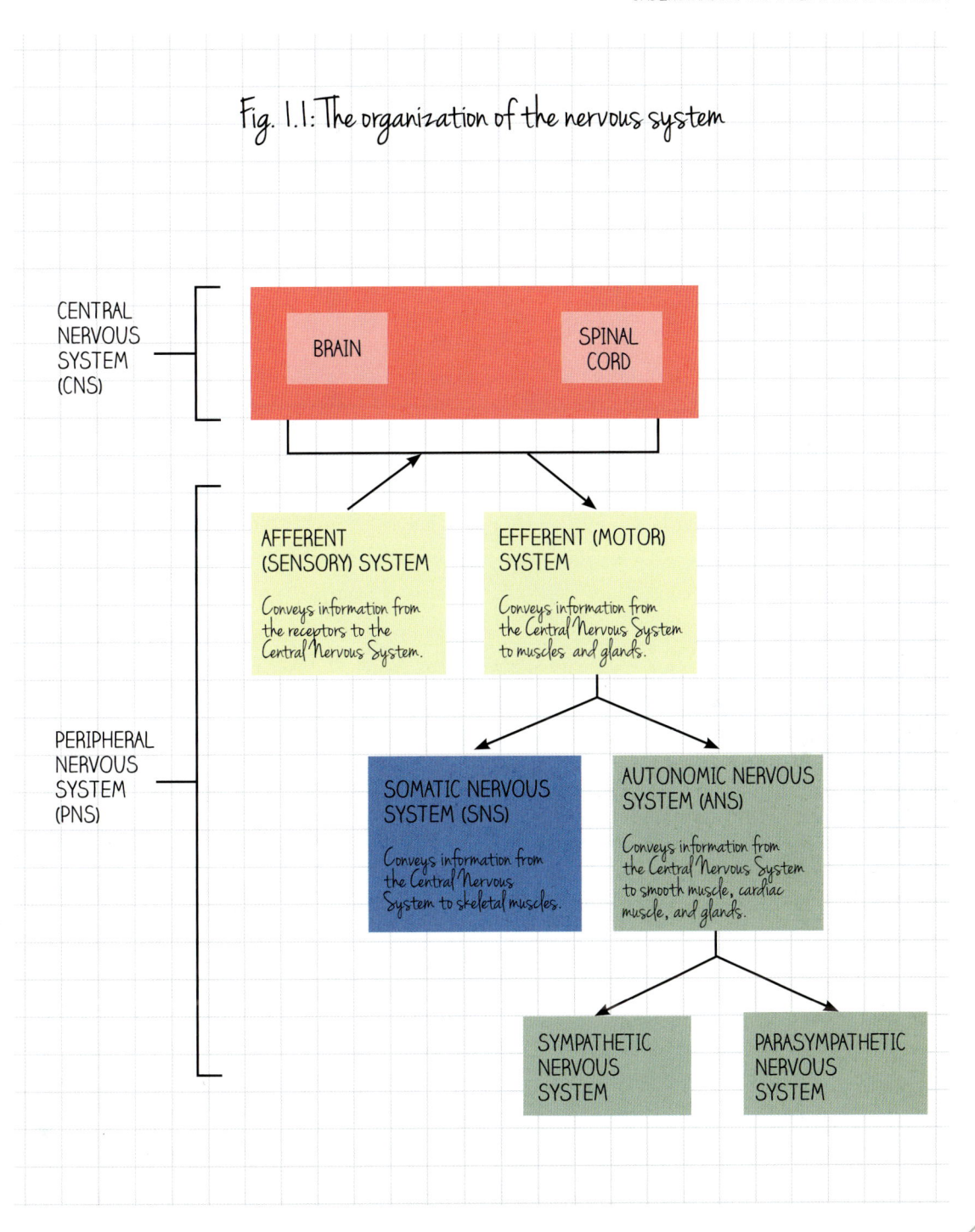

Fig. 1.1: The organization of the nervous system

CENTRAL NERVOUS SYSTEM (CNS)
- BRAIN
- SPINAL CORD

PERIPHERAL NERVOUS SYSTEM (PNS)

- **AFFERENT (SENSORY) SYSTEM**: Conveys information from the receptors to the Central Nervous System.
- **EFFERENT (MOTOR) SYSTEM**: Conveys information from the Central Nervous System to muscles and glands.
  - **SOMATIC NERVOUS SYSTEM (SNS)**: Conveys information from the Central Nervous System to skeletal muscles.
  - **AUTONOMIC NERVOUS SYSTEM (ANS)**: Conveys information from the Central Nervous System to smooth muscle, cardiac muscle, and glands.
    - SYMPATHETIC NERVOUS SYSTEM
    - PARASYMPATHETIC NERVOUS SYSTEM

ORGANIZATION OF THE NERVOUS SYSTEM

The Central Nervous System (CNS) comprises the brain and the spinal cord. The spinal cord, as its name suggests, runs from the base of the brain all the way down the centre of the spine, carrying information from each of our sensory receptors to all our muscles and glands. If this cord becomes broken or damaged, then the effects are devastating, because no information can pass from or to the rest of the body, leaving the injured person numb and in a paralyzed state.

The rest of the nervous system is described as the Peripheral Nervous System (PNS), and this includes all the nerves that carry information *in* from our sensory receptors – anything from touch and pain to levels of glucose and salt in the blood – as well as all the nerves that carry information *out* from the CNS to the muscles and glands. Many of the outgoing messages are under voluntary control – for example, those that go to the muscles allowing us to walk and talk. However, there is also a category of outgoing signals that relate to involuntary behaviours, such as the contractions of muscles in our intestines or hearts, the opening up of our lungs, or the release of saliva in our mouths. This branch of the PNS is called the Autonomic Nervous System (ANS) and its job is to balance the need to rest and digest versus the ability to respond to a "fight-or-flight" situation. As we will see in Chapter 4, this mechanism can sometimes come under strain, which has consequences for our physical and mental wellbeing.

INVESTIGATING THE BRAIN

How much do we actually know about the brain, though, and when did scientists and thinkers first begin to examine its precise role? As far back as 342 BC, the philosopher Aristotle highlighted the significance of the brain, proposing that the size of the brain in animals indicated the level of their intellectual ability. About five hundred years later, a famous Greek physician named Galen observed the effects

Right: The Peripheral Nervous System is made up of nerves that carry information to our muscles and glands.

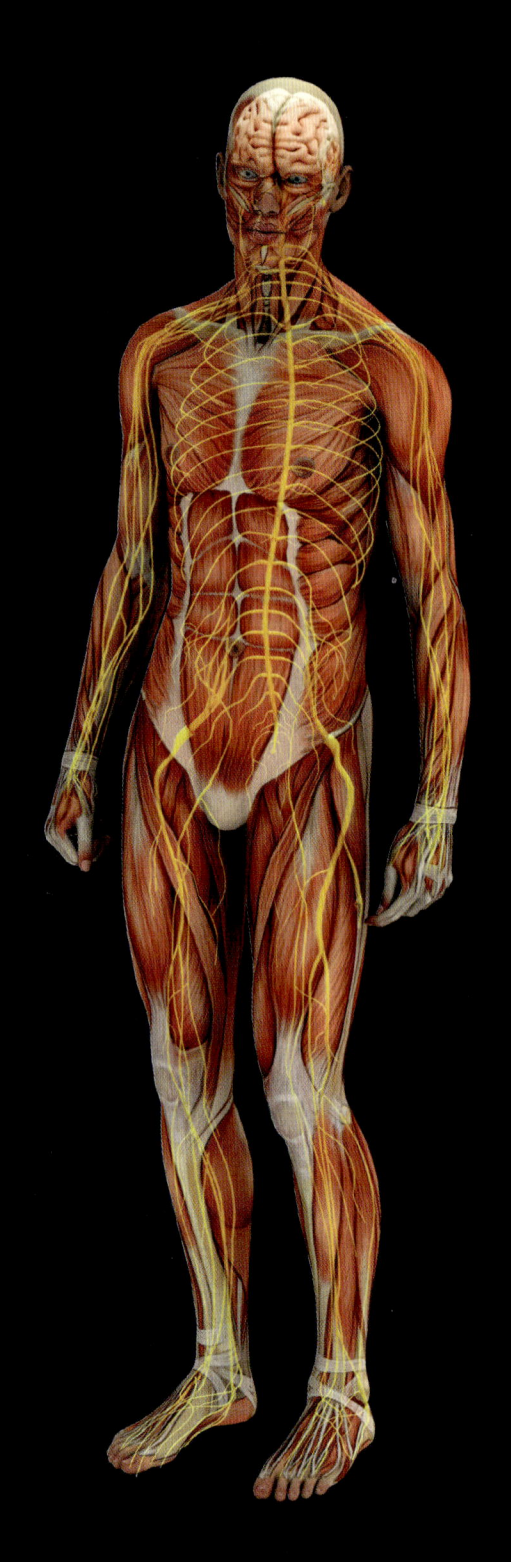

of brain injury in Roman gladiators and came to the correct conclusion that the brain controls our movements. So, our fascination with the brain goes back a long way, but it was not until the eighteenth century that scientists started to get a clearer idea of what the brain actually looked like and how it worked. By the nineteenth century, they knew something about the cells that made up the brain and had started to reliably identify which parts of the organ carried out specific functions.

Of course, all of the early scientific investigations relied on post-mortem examinations, as there was no way at that time of investigating the living brain. However, early in the twentieth century, a pioneering Canadian neurosurgeon named Wilder Penfield demonstrated that it was possible to evoke certain feelings and sensations by electrically stimulating different parts of the brain in conscious patients. Around the same time, the German psychiatrist Hans Berger invented the first electroencephalogram (EEG), which was able to measure the electrical activity of the brain. The latter part of the century saw an explosion of new techniques for observing the living brain – Computerized Tomography (CT), Positron Emission Tomography (PET) and Magnetic Resonance Imaging (MRI). Alongside the development of experimental psychology, these inventions led to incredibly sophisticated methods for unlocking the mysteries of how the brain creates the mind.

EARLY ATTEMPTS TO UNDERSTAND THE BRAIN

Back in the early nineteenth century, many people believed that you could accurately judge someone's personality, character and even their intelligence, simply by feeling the shape of their head. For example, someone with a protruding forehead was believed to have a benevolent nature, whereas a bump over the ear was thought to indicate that a person had destructive tendencies. The theory and practice of "phrenology", as this technique was known, was first put forward in the late eighteenth century and was based on work by the pioneering German neuroanatomist Franz Gall, who spent much of his life measuring the bumps, dips and ridges in the skulls of psychiatric patients, artists and criminals. His revolutionary idea was that the contours of the skull matched the shape of the underlying brain, and that the shape of the brain in turn reflected the strengths and weaknesses of an individual's character.

Gall's approach was to look for similarities and differences between the skulls of people with particular behaviours and/or talents. His painstaking research led him to identify a total of 27 fundamental characteristics, which he mapped onto a behavioural atlas of the skull. This "map" was then used as a kind of handbook: phrenologists were trained to feel carefully and measure the skull and would use this information to provide a detailed assessment of an individual's character and likely behaviour. The modern-day equivalent would be like a job interview where you were not asked to answer some questions and sit a few tests

Below: The German neuroanatomist Franz Joseph Gall (1758–1828).

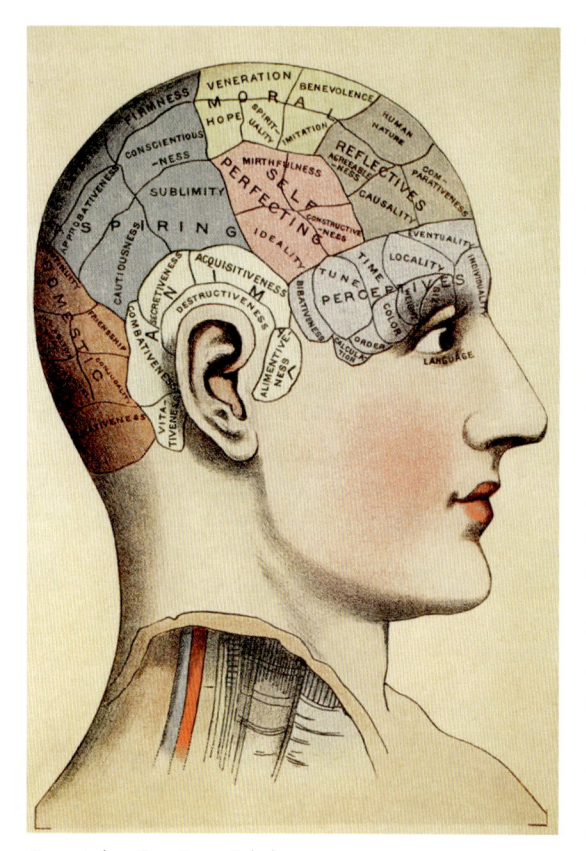

Above: A phrenological map of the human head, based on the work of Franz Gall.

but someone would be sent in to feel your head. In fact, because this technique was also used by psychiatrists in those days, this is where the expression, *You need to get your head examined!* comes from.

Although many of Gall's conclusions were wrong, his work is often seen as an important landmark in the development of modern brain research. He proposed that the brain was an amalgamation of lots of mini-organs, each of which represented a specific trait, such as self-esteem or secretiveness. While his list of traits might seem rather laughable to us now, the basic premise that different parts of the brain perform different roles has become widely accepted. While Franz Gall was not the first to suggest this link between brain structure and function, he certainly brought these ideas to the fore. Importantly, he maintained that the stronger a particular trait, the bigger that part of the brain, hence causing a bulge in the relevant part of the skull. Once again, Gall was wrong – we now know that the shape of the skull does not tell us anything much about the brain inside it. However, he was certainly thinking along the right lines when he suggested that the brain is shaped by our behaviour, experience and environment.

In some ways, the most radical stride that phrenology made was proposing a method for assessing the living brain. As we know, the only way to examine a brain up until this point was to wait for someone to die and then carry out a post-mortem – or, as one prominent phrenologist said, *All we knew about the brain was how to slice it.* Franz Gall may have been wrong in thinking that the shape of the skull gave us a window into the brain, but his research was an important step forward and paved the way for others to come up with better ideas.

MODERN BRAIN IMAGING

In fact, it took more than another hundred years to pass before scientists could really "see" inside the brain in any reliable way. Nowadays, of course, there are many advanced technologies that have led to huge leaps in our knowledge about how the brain works, which bits do what, and what happens when it goes wrong. In fact, in our modern world the choice of neuroimaging procedures can seem rather overwhelming, even for those who work as neuroscientists, psychologists or doctors!

In this section we will have a brief look at some of the ways that we can get a window on the brain – but because there are a number of different approaches, it is useful to pause and consider what questions scientists are trying to answer. Firstly, what does the human brain actually look like: what is its precise and detailed anatomy? And which areas of the brain correspond to different functions – for example, which parts are involved in our ability to do maths, or the control of our sleeping and waking? For clinicians, it can be helpful for diagnosis and treatment if they know which areas of the brain are damaged or function differently in a person who presents with a particular disease or psychiatric disorder, such as dementia, schizophrenia, Parkinson's disease or autism.

EXAMINING BRAIN STRUCTURE

When it comes to examining the structure of the brain, the good old-fashioned post-mortem still offers a huge amount of insight that we cannot glean from even the very best neuroimaging. Up-to-date techniques mean that the brain of a deceased person can be cut into extremely thin slices, and histologists (cell and tissue analysts) can use special stains to reveal incredible detail about all the cells and other structures within sections of the brain. This allows scientists to establish the precise cellular structure of various areas of the typical brain, but the same approach can also be used to show up differences in the brains of people with a particular disorder. For example, despite the fact that Alzheimer's disease is very prevalent, it remains true that we can give a definitive diagnosis of this condition only after someone dies. It is only then that histologists can take slices of the brain and identify the increased presence of *plaques* and *tangles*, the tell-tale signs of this cruel degenerative disease.

Of course, post-mortems are not much use if we want to look at the structure of the living brain, so Computerized Axial Tomography (CT/CAT) and Magnetic Resonance Imaging (MRI) offer a practical alternative. CT technology was the first of these techniques to be developed, becoming available around the early 1970s. CT scanners essentially produce a computerized composite of many x-rays. During the process, an individual lies inside a tunnel-like machine and x-ray beams and detectors rotate around the head, collecting multiple images. The computer then pieces these together to create an overall picture of the brain. These scans offer the option to see the brain from many different planes and even in 3D. CT scans are frequently used to detect signs of stroke, haemorrhages and tumours, and offer a relatively inexpensive and widespread option for brain analysis.

Below: Cutting the brain into very thin slices allows scientists to examine its structure in incredible detail.

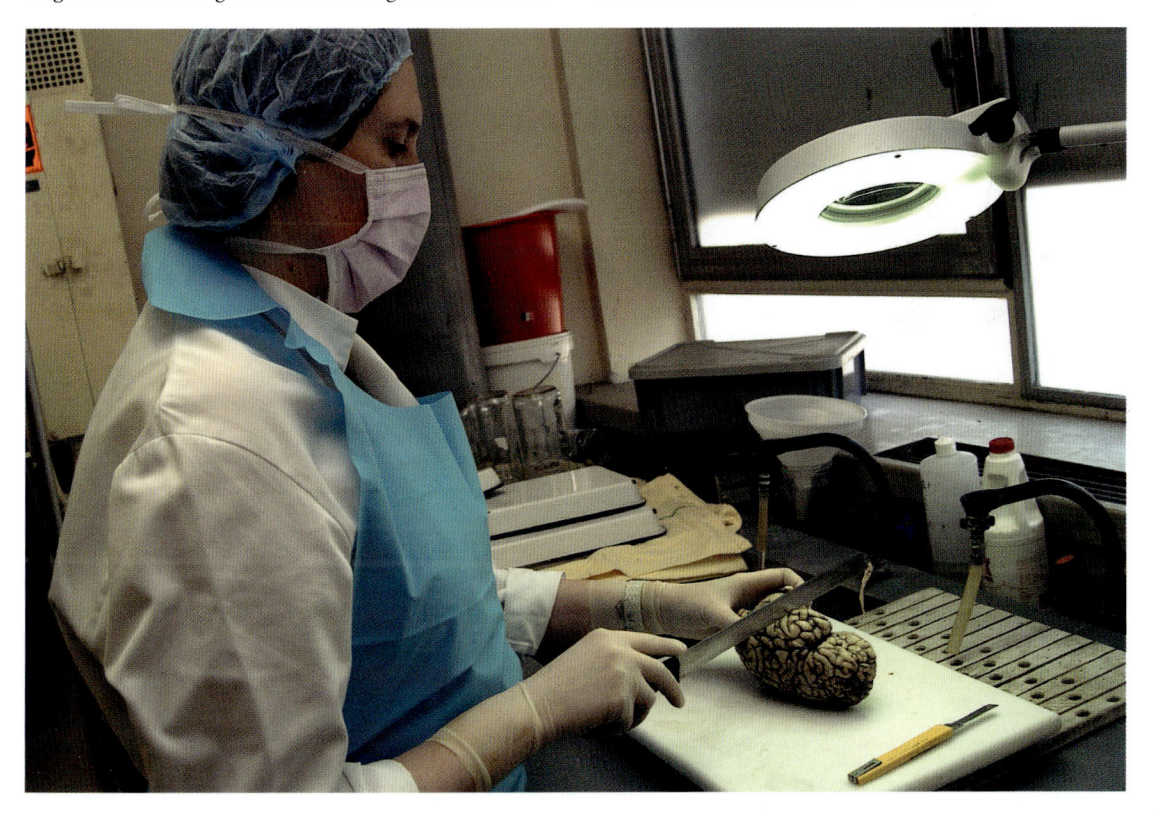

UNDERSTANDING AND INVESTIGATING THE BRAIN

The downside of CT scans is that they expose the individual to ionizing radiation and have the capacity only to produce images of limited resolution. This means that a CT scan cannot provide the same level of detail as an MRI scan, a technique that came into commercial use about ten years later. The procedure for an MRI scan is a similar experience for the individual being assessed. However, instead of using x-rays, this technology employs huge magnets to spin the hydrogen atoms in the brain. This in turn causes the hydrogen atoms to emit a detectable signal, which provides very meticulous data regarding water (H_2O) distribution and density. This is an ingenious way of looking at soft tissue and provides an incredibly detailed image of the structures in the brain (*see* below).

The concept of how an MRI scanner works might sound rather frightening, but although the procedure involved is indeed quite noisy and slightly claustrophobic, it is in fact extremely safe. That said, although an MRI scan is safer and far more detailed than a CT scan, it is significantly more expensive to conduct. The procedure involved also takes longer to carry out, which can be a practical disadvantage, not least because it is difficult to keep patients lying very still for longer periods of time.

USING THE DAMAGED BRAIN TO UNDERSTAND THE LINK BETWEEN STRUCTURE AND FUNCTION

The scanning techniques outlined above tell us a lot about the anatomy of the brain, but on their own say little about which part does what. After Franz Gall initially proposed the theory of phrenology in the late eighteenth century, a number of physicians and scientists continued the quest to find out exactly where different cognitive and behavioural functions are located in the brain. One of the biggest breakthroughs came from Paul Broca, a French physician who practised in the mid-nineteenth century and who is credited with discovering the speech centre in the brain. Broca had a patient named "Tan", so-called because this was the only word he could articulate. Tan had a long history of progressive loss of speech and, by the time Broca met

Below: On the left is an image of the brain produced by a CT scan. On the right, the MRI scan provides a far more detailed picture.

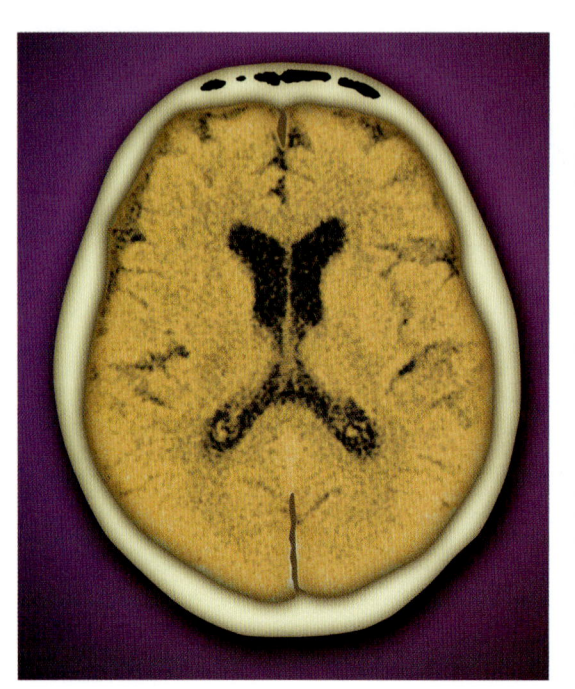

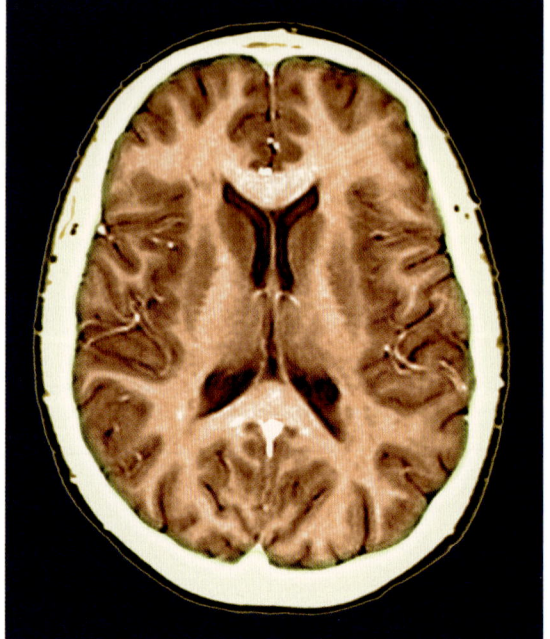

him, had been unable to speak for nearly 21 years. However, he appeared to be otherwise intellectually quite capable, seemed to understand what was being said to him and still made every effort to communicate.

Broca was extremely fascinated by this patient, so when Tan eventually died of gangrene on 17 April 1861, the quick-thinking doctor hurried straight along to the morgue to try and find out what might be different about this man's brain. An autopsy revealed a specific lesion – an area of dead tissue – in Tan's frontal lobes, leading Broca to conclude that this area was vital for speech production. He followed up this case by examining twelve more individuals who showed speech difficulties, and was soon able to confirm and extend his original findings. Although others had done similar work a generation before, their research was never fully written up, so it is this series of case studies by Broca that is generally seen as paving the way for the surge of subsequent research into how language is produced, received and understood by the brain. To this day, the specific region of the brain that controls speech production is called *Broca's area* and people who are unable to produce speech are described as having *Broca's aphasia*.

There are two good reasons why language was the first function to be localized in the brain. The first is that the comprehension and production of speech are both very easy things to measure, and the second that most humans reach a reasonably similar level of competence in speech by the time they reach adulthood. It is therefore a relatively simple step to consider which elements of language might not be working properly and then to use a post-mortem, or in more recent times an MRI or CT scan, to see which part of the brain is damaged. We then infer that the damaged part of the brain is normally responsible for the function that has become lost. To give an analogy, it may be possible to work out how a vehicle braking system works by finding a car that will not stop and then comparing it part-by-part with one that does.

This practice is called *neuropsychology*, and it has been used to localize many other functions, including face recognition. One vivid example of neuropsychology is the case of the farmer who could no longer recognize the faces of any of his family but had no problem naming his cows. This suggested to neuropsychologists that these two processes might be handled separately in the brain. To be absolutely sure of this, it was necessary to find someone who had the opposite problem. Sure enough, a later case study described a shepherd who, after injury, was able to name all of his family but could no longer name his sheep. This kind of dissociation has also been found with language – there are people who can understand speech but cannot produce it, and others who can produce speech fluently but cannot understand what someone is saying to them. This suggests that speech production and speech perception are dealt with separately in the brain.

As well as being a fantastic research tool, neuropsychology can be used very effectively in a clinical setting. We can pool together the findings from the past 180 years or so and reverse the concept: there is now a standard set of cognitive assessments that measures skills such as language, memory, attention, planning, and control. By analyzing a person's performance across these tests, an experienced neuropsychologist can have a good guess at which part of

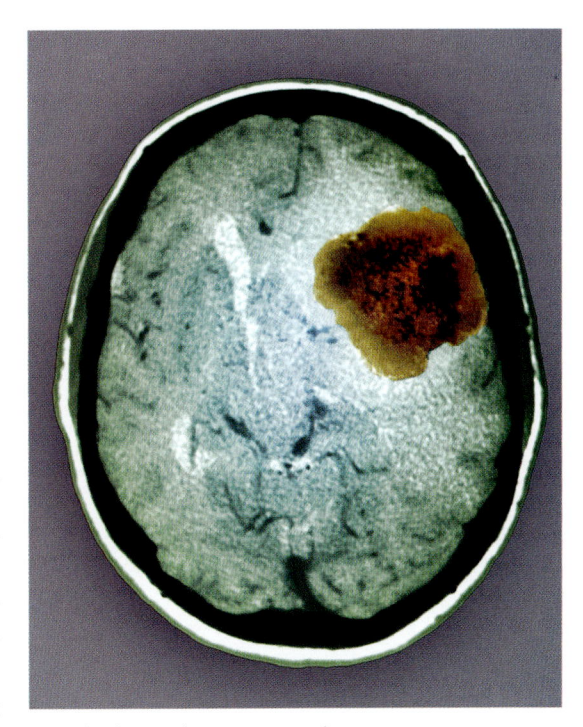

Above: MRI scans play an important role in diagnosing brain tumours.

UNDERSTANDING AND INVESTIGATING THE BRAIN

the brain might not be functioning properly. This plays an important part in diagnosis and further investigation. For example, an older patient who shows a specific impairment of memory but who does well on other tests may be in the first stages of Alzheimer's disease, or may have a tumour in the medial temporal lobes.

WATCHING THE BRAIN IN ACTION

Neuropsychology is one method of localizing different functions in the brain, but there are also various imaging technologies that can add to our knowledge. The electroencephalography (EEG) technique first put forward by Hans Berger (*see* page 13) is still a much used and respected method with which to assess brain function. Electrodes placed on different areas of the scalp can record how excited the neurons are and the extent to which they are firing together. These in turn indicate what type of activity that area of the brain is engaged in. Beta waves, for example, are seen when a person is carrying out a focused, analytical task, whereas alpha waves are usually present during periods of relaxation and creativity. Delta waves, which are much bigger and slower, suggest deep sleep. Because EEGs are very accurate records of timing, researchers can use them to pinpoint particular moments of thought or action very precisely.

The disadvantage of EEG is that it does not localize very accurately. When using the technique, we may know that activity happens somewhere in the frontal lobe, but not necessarily where. EEG is also unable to pick up activity deep in the brain. However, there are two very good alternatives. The first of these is Positron Emission Tomography (PET), which involves injecting a person with *labelled glucose* – that is, glucose that gives off radiation that can be traced. A scanning machine is then used to detect areas of radiation in the individual while they carry out a specific task, such as trying to remember some specific words. The basic principle of PET is that the active regions of the brain will be using the glucose, so by tracking it we can see which parts are working hardest. Again, this sounds a little frightening at first, but the labelled glucose is in fact a harmless, naturally occurring substance and the technique is actually very safe. Possibly the most widely used imaging technique these days is functional MRI (fMRI), which combines all the clarity of a standard structural MRI scan with the advantages of seeing which parts of the brain are active. This method assesses blood flow, which is generally thought to be a good indication of activity. Because it is so intriguing to be able to pinpoint accurately which areas of the brain are firing when people are thinking or doing particular things, studies that use this technique often get a lot of publicity. The two biggest news headlines currently relating to it are that fMRI can tell you how intelligent you are and can predict the efficacy of anti-smoking advertisements. This technique has been used to do many things, ranging from evaluating an individual's political views to examining the brain activity of a woman during orgasm. The point is that fMRI is extremely appealing

Above left: Cognitive assessments measure skills such as memory, attention and control.

Opposite: The EEG electrodes (above) provide a read-out (below) which shows typical brain waves.

UNDERSTANDING AND INVESTIGATING THE BRAIN 19

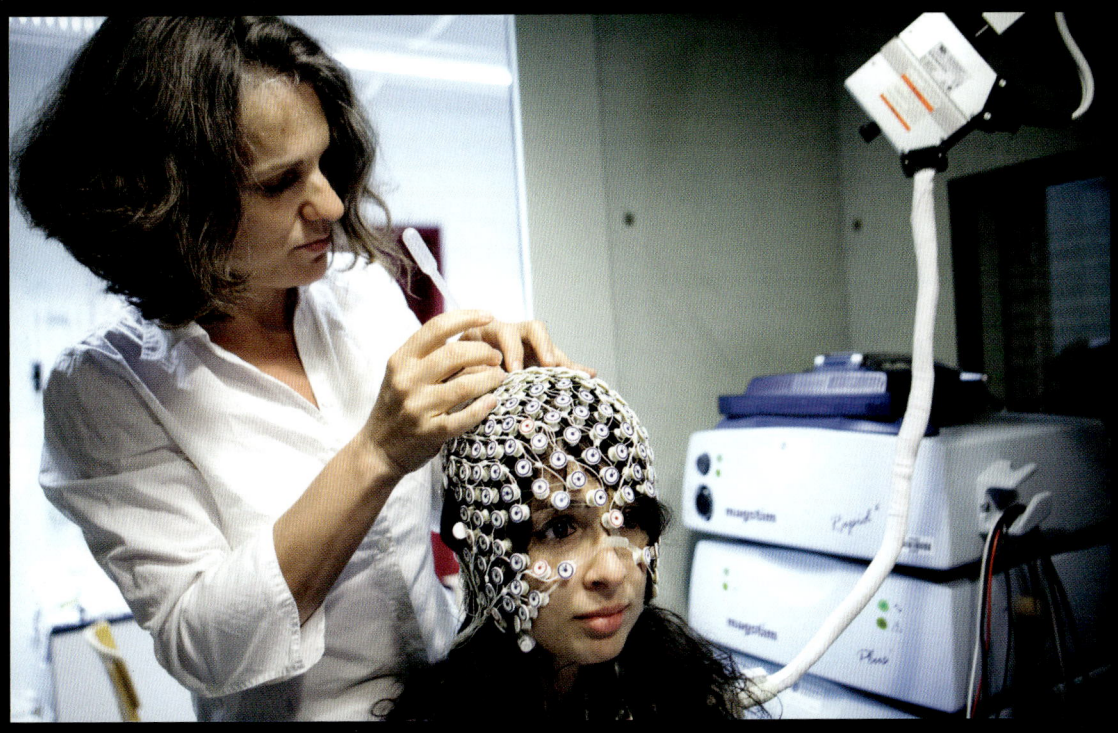

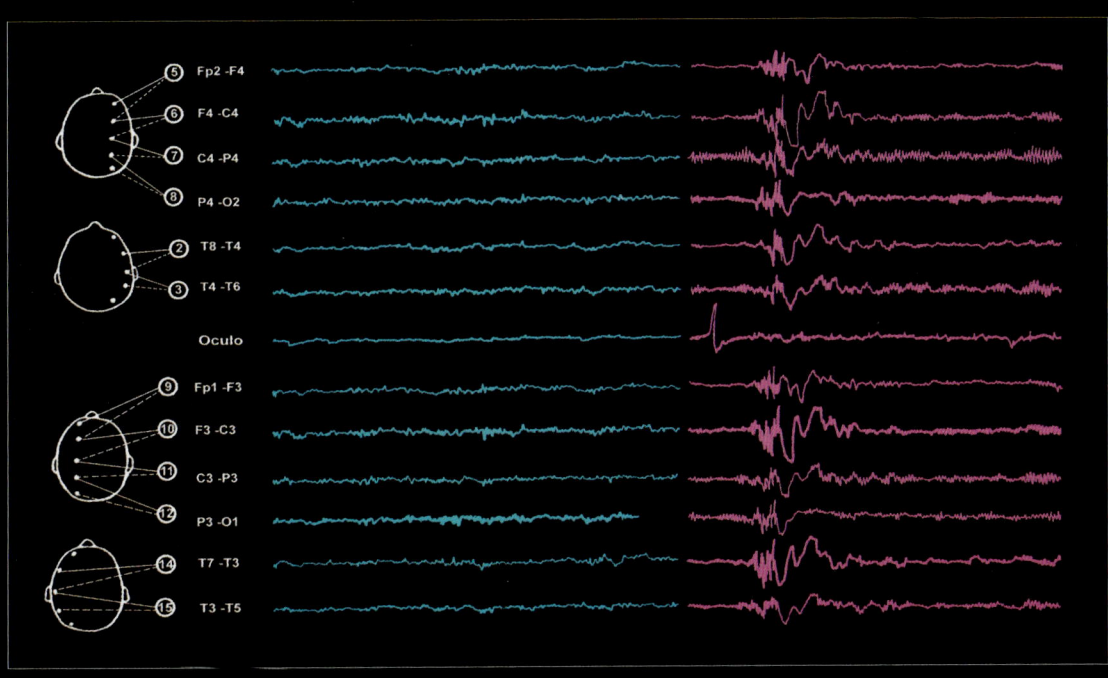

20 UNDERSTANDING AND INVESTIGATING THE BRAIN

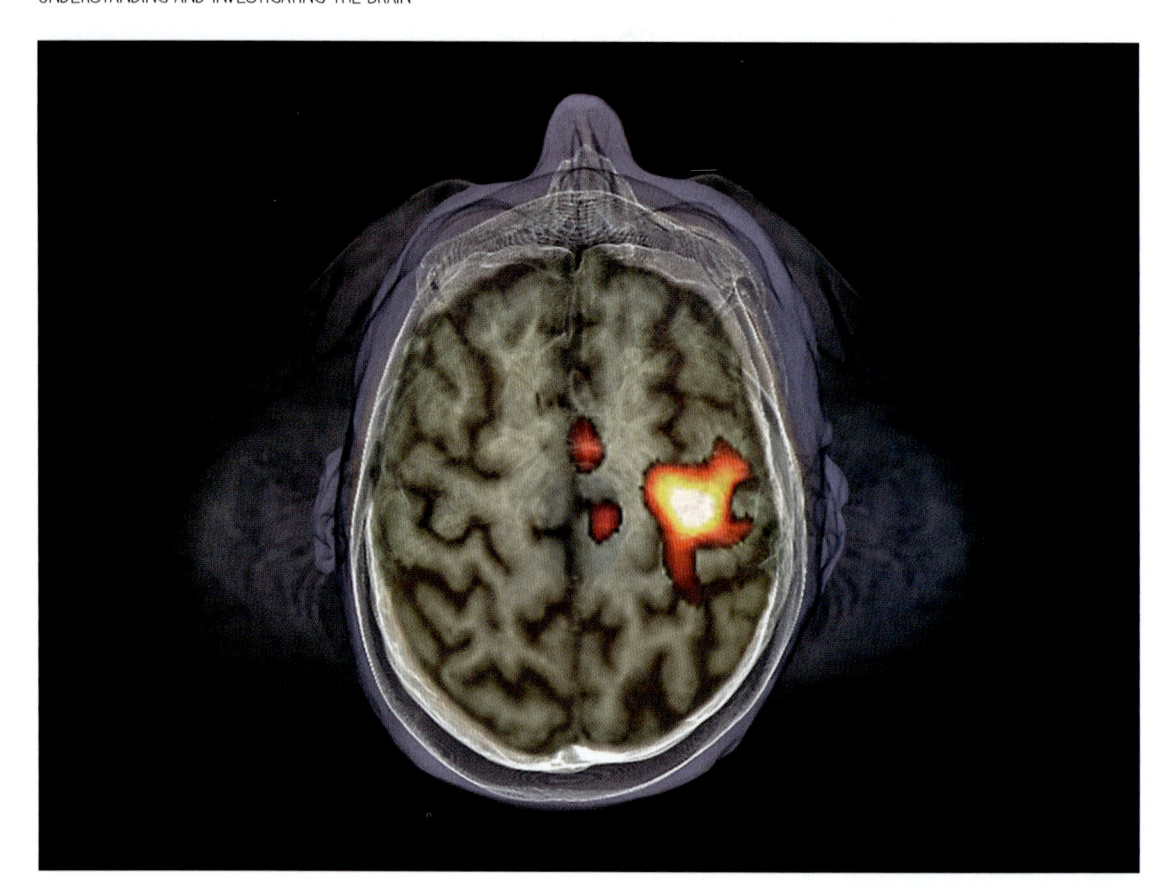

and generally a very good technique when used and analyzed properly. However, that does not mean that it is not sometimes misused, so it is important to view findings with a degree of caution and not to be blindly seduced by them.

A recently developed technique that also looks at brain function, but in a totally different way, is Transcranial Magnetic Stimulation (TMS). This uses magnets to temporarily increase or decrease activity in particular regions of the brain. Some research might use this approach to cause a short-term loss of activity in a particular area, which is another way of establishing what that area of the brain does. But the method is also used to stimulate brain activity and has been deployed clinically for a number of medical conditions, including migraines, strokes and depression. Again, there is some controversy around the manner in which it is used, especially in young children, and some have issued

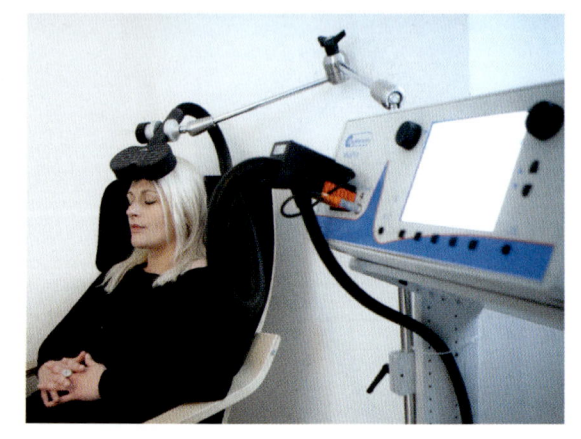

Top: fMRI scans enable us to determine which parts of the brain are active.

Above: TMS has been used clinically for migraines, strokes and depression.

strong warnings against Internet websites that teach people how to make their own brain stimulator. Nevertheless, TMS is without doubt a very promising new approach to understanding and modifying brain activity.

WHAT DOES ALL THIS RESEARCH TELL US?

Thanks to all of this varied investigation, we now have a fairly good idea of what the brain looks like and what the different regions of the brain do. At the bottom and middle of the brain, at the junction with the spinal cord, is the *brain stem*. This is the main relay station for all traffic coming in and out of the brain and is also responsible for many of our vital functions and reflexes, including heartbeat, breathing, swallowing, coughing and sneezing. It also plays a key role in consciousness and sexual arousal.

Just above the brain stem are found the *thalamus* and *hypothalamus*. The thalamus is sometimes described as a "multi-media mixing console", in that it relays and integrates information from each of the senses. It is also involved in the interpretation of pain and temperature, as well as playing a role in memory and emotion. The hypothalamus, found just below the thalamus, is a very small part of the brain – around one per cent of its total mass – and yet it is absolutely vital to maintaining the correct balance of oxygen, glucose, temperature, hydration and so on. It also controls the pituitary gland, which in turn regulates all of our hormones. Quite simply, this very small part of the brain keeps us alive and well.

At the very back of the brain is a large cauliflower-shaped piece, which is called the *cerebellum*. It is this area of the brain that records and stores all of our motor learning – that is, the ability to walk, ride a bicycle, play tennis, play the piano and so on – as well as helping us to maintain balance and posture. The cerebellum also appears to be very important in

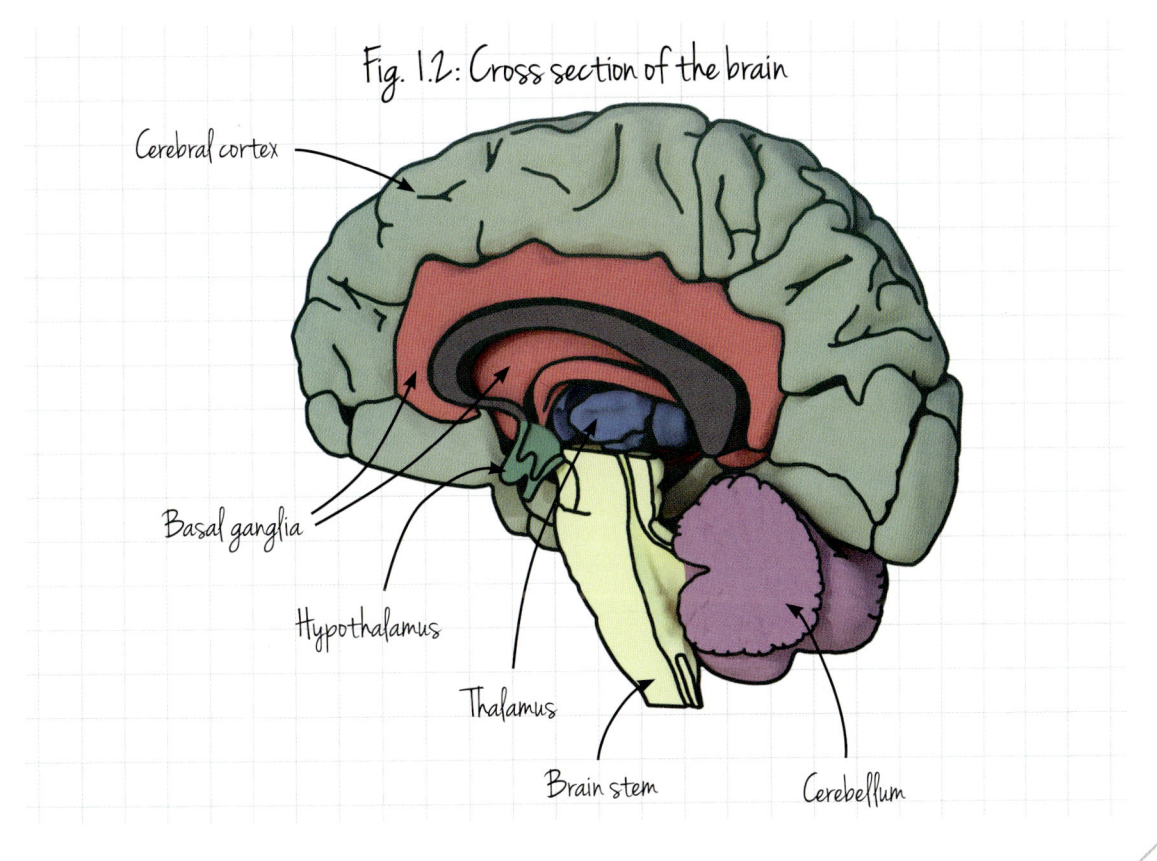

Fig. 1.2: Cross section of the brain

- Cerebral cortex
- Basal ganglia
- Hypothalamus
- Thalamus
- Brain stem
- Cerebellum

UNDERSTANDING AND INVESTIGATING THE BRAIN

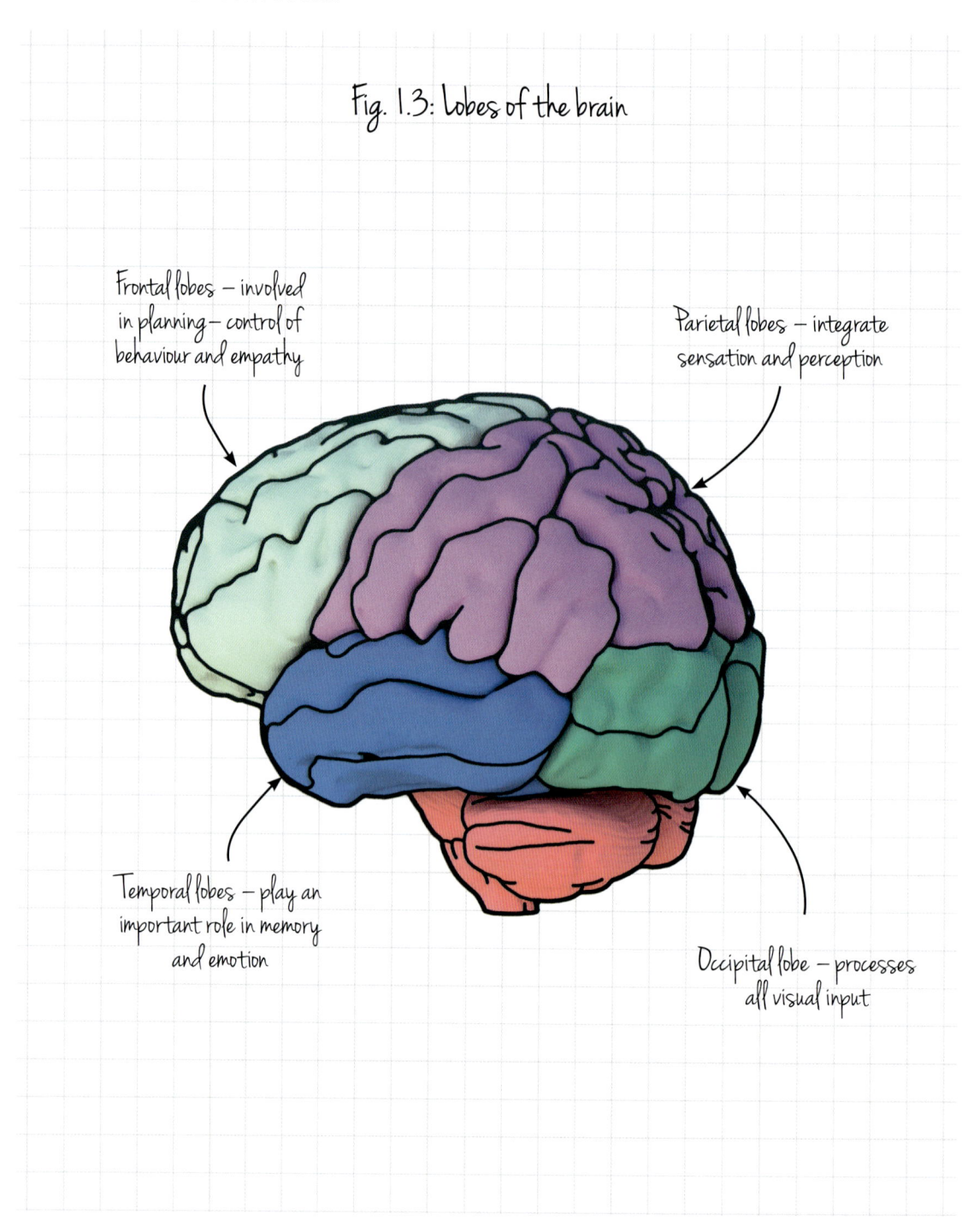

Fig. 1.3: Lobes of the brain

Frontal lobes — involved in planning — control of behaviour and empathy

Parietal lobes — integrate sensation and perception

Temporal lobes — play an important role in memory and emotion

Occipital lobe — processes all visual input

managing our emotions and our capacity to pay attention, partly because it has many long-range connections with the frontal lobe, at the opposite end of the brain.

The remaining part of the brain is described as the *cerebrum*, which is divided into two hemispheres, both roughly symmetrical. Toward the middle of this area lie the *limbic system* and *basal ganglia*. Our emotions, sensations of pleasure and pain, and memory formation are all very much driven by the limbic system. The basal ganglia are strongly interconnected with many other important structures and seem to play an important role in regulating movements and thoughts. Damage to circuits containing the basal ganglia have been linked with many medical conditions, including Tourette's syndrome, Obsessive Compulsive Disorder (OCD) and depression.

Finally, we come to the cerebral cortex, the outer layer of the brain, which is 2–4 mm (0.07–0.15 in) thick and heavily folded in order to increase surface area. This area is usually described in terms of four different kinds of lobes – the frontal lobes, the parietal lobes, temporal lobes and the occipital lobes. The cerebral cortex can be thought of as the intelligent, thinking part of the brain, managing many of our sensations and perceptions, memories, thoughts and behaviours. The frontal lobes in particular seem to be vital to understanding other people, controlling our behaviour, planning our actions and providing a mental working space. They have been greatly studied over the last few decades and are often described as the seat of our intelligence and the location of our personalities. This part of the brain is larger in humans than most other animals and is also the region that develops last when the brain is still growing. Amongst other things, this explains the behaviour associated with the Terrible Twos, the term widely used to describe very young children who seemingly act without reason or control of any kind. Conversely, the frontal lobes are also the part of the brain that ages fastest.

Over the past century or so, we have learnt so much about how the brain makes us who we are – but there is also still so much more for us to learn. Indeed, it is entirely possible that we may never get to the bottom of some aspects of how the brain works; for example, the great puzzle of what creates consciousness. The next decade will doubtless prove to be an exciting time, as research takes us ever further into the mystery of the brain and how it makes us who we are.

CASE STUDY: "NICK"

"Nick" was a 35-year-old man with a job as a CEO of a large company. One evening he was cycling home when a large twig got caught up in his front wheel, causing him to fly over the top of his handlebars. He landed headfirst and, although his helmet took the brunt of the impact (and almost certainly saved his life), he lost consciousness and was taken to hospital. Nick made a quick recovery from his physical injuries, but he and his family were aware that he seemed to be a different person. He was no longer able to concentrate for any length of time and found it difficult to organize himself. Having always been a very calm person, Nick became far more irritable and started to have dramatic mood swings. He developed a need for everything to be in a particular order and became very anxious when his plans were disrupted. A scan showed that the accident had injured the front of his brain – the prefrontal cortex. Unfortunately, the brain does not repair itself in the same way that a broken bone does and so Nick has had to accept that he will no longer be quite the same person he was before. However, although doctors were not able to offer a cure, they were able to refer Nick to a neuropsychologist for cognitive rehabilitation and emotional support (*see* also page 186). This helped Nick to adjust to his new self and compensate for some of his lost skills so that he was able to return to work.

2. HOW DOES THE BRAIN WORK?

WHAT IS THE BRAIN MADE FROM? THE BUILDING BLOCKS OF THE BRAIN ARE TWO MAIN TYPES OF CELLS: NEURONS AND GLIAL CELLS.

Neurons are specialized cells that are able to carry messages in the form of electrical energy. It is estimated that there are around one hundred billion of these cells in the average human brain, which is more than fourteen times the population of the world! There are at least as many glial cells as there are neurons. These provide vital support and protection for the neurons, but can be affected by certain diseases, such as multiple sclerosis, and are a common source of tumours in the nervous system. Recent findings have suggested that these cells may also play a role in the way neurons communicate with each other.

NEURONS

Neurons display a number of important characteristics that equip them for carrying messages around the body. At one end is a cell body, which contains all the genetic information and energy stores. The cell body has lots of branch-like extensions called *dendrites*, which act as receivers, collecting information from other surrounding neurons. In addition to these, there is one very long extension – an *axon* – that is

Below: Glial cells provide support and protection for neurons in the Central Nervous System.

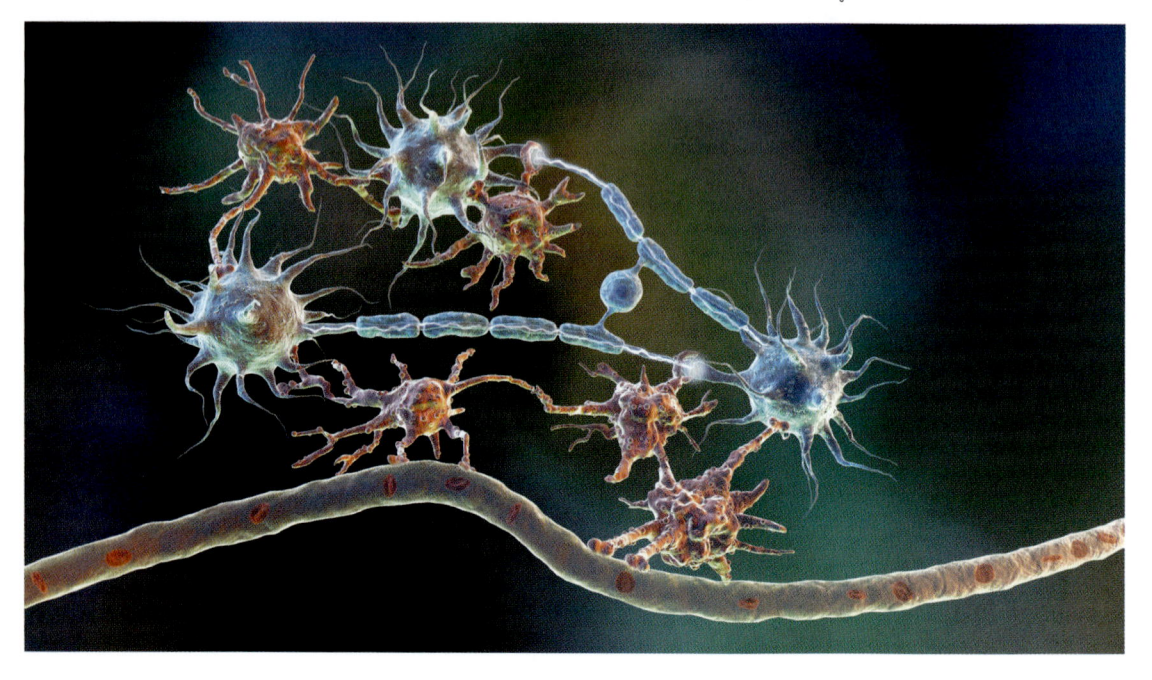

like a one-way street heading away from the main cell body. The axon is a route, along which the signals can be carried to other neurons and other parts of the body. The longest neuron in the body goes from the base of the spine to the big toe and can be around 1 metre (1 yard) long, which is very big indeed for a single cell.

We can classify neurons by the job they do. Sensory neurons collect information from the external and internal environment – they detect everything from touch and smell to pain and oxygen levels. Motor neurons carry signals from the brain to our muscles and glands, so these include those that instruct our biceps to contract when we want to pick something up, or cause our mouth to water when we smell something nice. Finally, there are many interneurons, which control and modify the communication between the sensory and motor neurons.

HOW DO NEURONS CARRY ELECTRICAL SIGNALS?

One of the most intriguing things about a neuron is its ability to transmit electrical signals; and we have the giant squid to thank for much of our current understanding of how this happens. This monster of the sea, which can grow up to 13 metres (43 feet) in length, has an axon that is much larger in diameter than that of most other animals. Because of this, it is possible to use electrodes to measure exactly what is happening when a neuron is at rest and when it is firing. A series of important experiments was begun in the 1930s by the English physiologists and biophysicists Alan Hodgkin and Andrew Huxley, and over the next 25 years or so they were able to pin down the precise mechanisms by which neurons communicate. In fact, so influential was their work that in 1963, together with Sir John Eccles, they were awarded the Nobel Prize for the Hodgkin-Huxley model of neuronal function.

In order to get to grips with Hodgkin and Huxley's findings, we need to consider some basic chemistry. Within the neuron, and indeed all around us, there are many atoms and molecules that carry an electrical charge. These are called *ions*. To give a specific example, table salt is also known as sodium chloride. When it dissolves in water it separates into sodium ions, which carry a positive charge, and chloride ions, which carry a negative charge. As it happens,

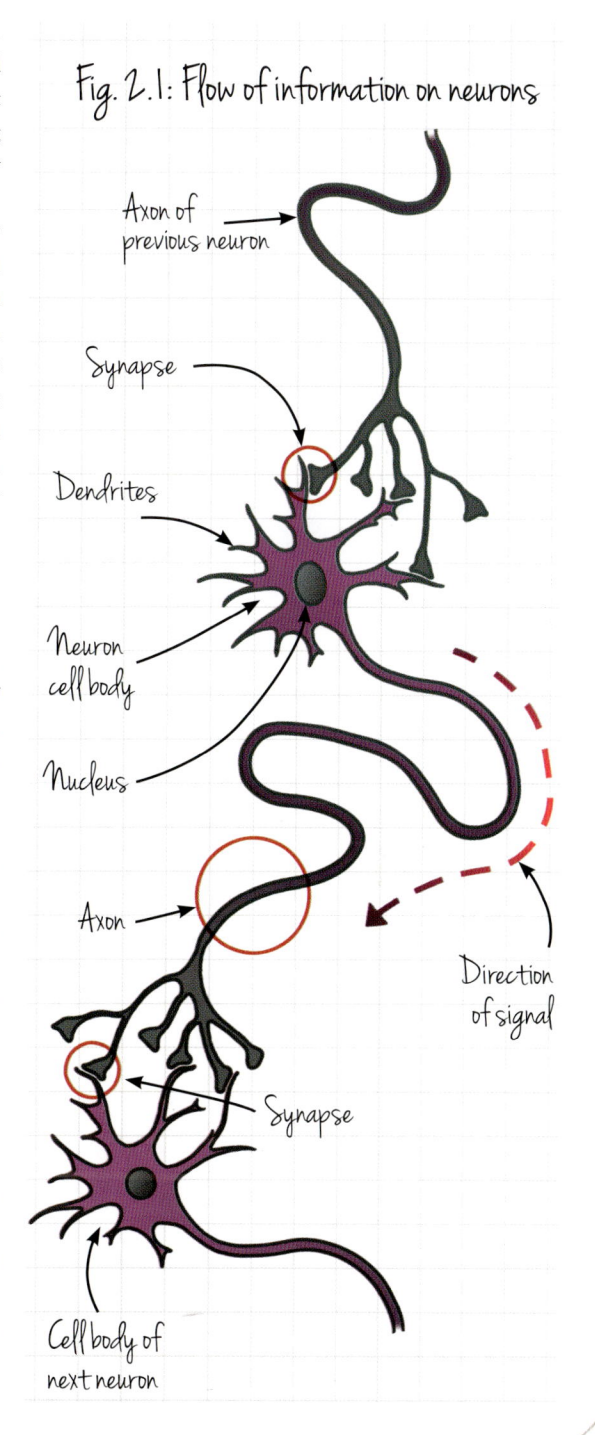

Fig. 2.1: Flow of information on neurons

Above: The cell membrane is made up of two fatty layers interspersed with gateways called ion channels.

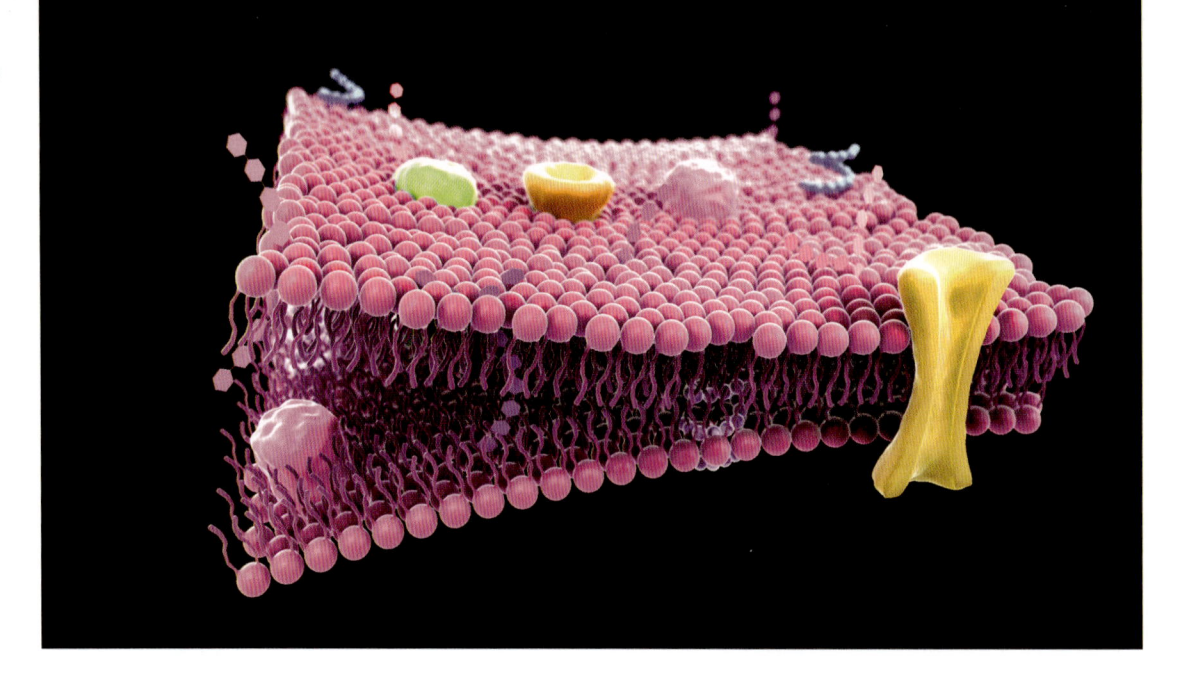

these two particular ions, which are found everywhere in our environment, play a leading role in carrying signals in the brain. Another ion that is particularly important is potassium, which also carries a positive charge.

So, ions are key players in transferring messages in the brain, but to understand how they do this we have to look at how they move in and out of the cell. Around the outside of the neuron is a cell membrane, which is made up of two fatty layers, scattered with receptors and interspersed with special gateways called ion channels (*see* image above). There are different types of ion channels, which are like guards that determine what can pass across the membrane and when. Ions can only pass through particular channels and, to give an extra layer of control, many of these channels have "gates" that can be opened or closed with changing circumstances. As we will see, controlling the movement of these electrically charged ions is crucial to the way in which the electrochemical signal is transmitted.

THE NEURON AT REST

First of all, let us establish what is happening in a neuron when it is at rest – that is, not firing. The early experiments by Hodgkin and Huxley revealed something very important about a resting giant squid axon: the inside of it was electrically negative compared with the outside. Specifically, they showed that there was an electrical charge across the membrane of around -65 millivolts (mV). This is described as the *resting potential*, and when a neuron is in this state it is ready and able to fire, as soon as it receives the right level of stimulation.

Additional experiments showed that the resting potential occurs because of an unequal distribution of ions (you will remember these carry electrical charges). Fig 2.2. (opposite) illustrates that there are varying amounts of sodium (+), potassium (+), chloride (-) and organic anions (-) either side of the membrane – for example, there is a lot more potassium inside the cell than outside. These ions are continually moving around, but while the neuron is at rest, the concentrations remain quite stable and, on balance, there are always more negative ions inside the cell than outside it, hence the negative charge.

This resting potential is absolutely vital. If it is wiped out – as might happen, for example, when the brain is starved of oxygen – the neuron is no longer able to fire. So how does the neuron maintain this state of readiness? From basic physics, we know that there are two natural forces that pull these ions backward and forward across the cell in a kind of continual tug-of-war. The first of these is *electrostatic*

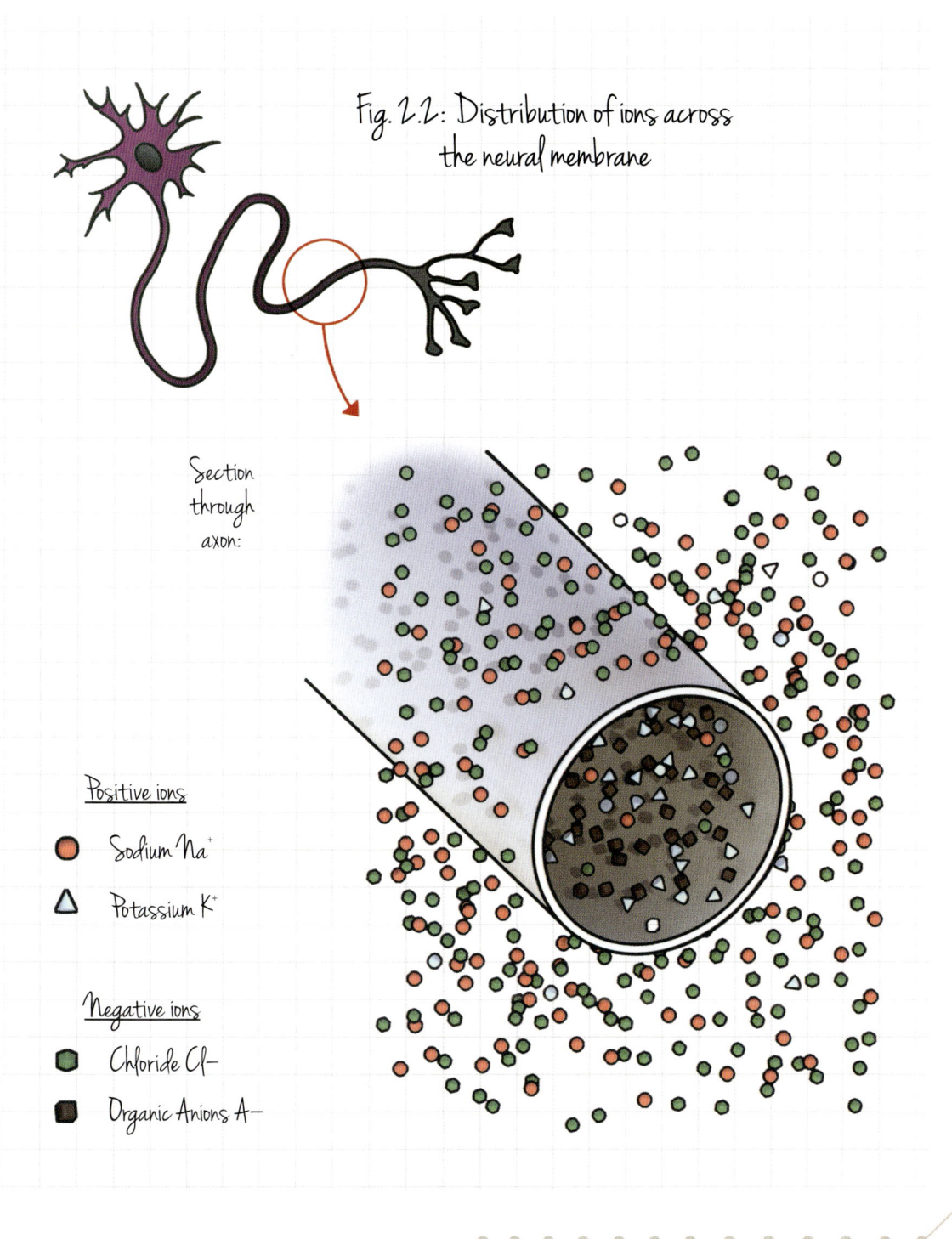

HOW DOES THE BRAIN WORK? 27

Fig. 2.2: Distribution of ions across the neural membrane

Section through axon:

Positive ions
- Sodium Na$^+$
- Potassium K$^+$

Negative ions
- Chloride Cl$^-$
- Organic Anions A$^-$

Fig. 2.3: Action potential

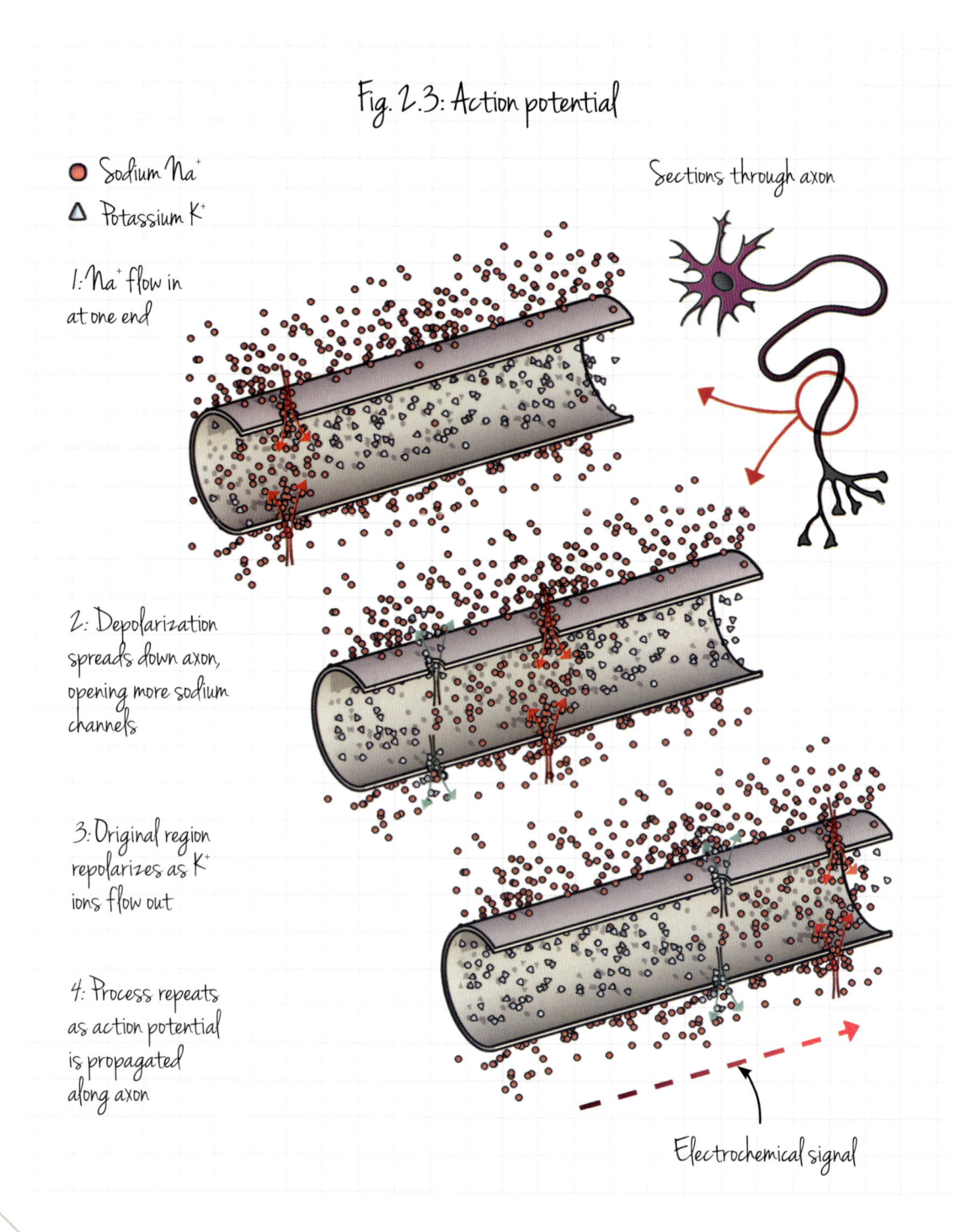

- ● Sodium Na⁺
- △ Potassium K⁺

Sections through axon

1: Na⁺ flow in at one end

2: Depolarization spreads down axon, opening more sodium channels

3: Original region repolarizes as K⁺ ions flow out

4: Process repeats as action potential is propagated along axon

Electrochemical signal

pressure, a force that drives positive ions toward negative ions and vice versa. The second force, *diffusion*, is the tendency of any substance to spread itself out – that is, to move from where it is very concentrated to where it is less concentrated. A simple example to demonstrate this principle is that if you put something strong-smelling in one corner of the room, the aroma will gradually drift away from that concentrated spot and spread itself out across the room. The same is true of ions – they will tend to move from where there are lots of them to where there are fewer.

When the neuron is at rest, these two forces are often acting against each other. If we look at potassium, for example, diffusion is drawing the ions out of the cell to where the concentration is lower but, because it is a positive ion, the force of electrostatic pressure is enticing it back into the cell, where there are excess negative ions. With chloride, the principle is the same but reversed. However, for Hodgkin and Huxley, the real puzzle was what was happening with the sodium ions. In this case, both diffusion and electrostatic pressure act in the same direction – that is, pulling *into* the cell. Nevertheless, the concentration outside remains much higher than inside.

A potential explanation for this would be that the membrane is impermeable to sodium ions – they simply cannot cross. This seemed quite likely at first, because although there are lots of sodium channels in the membrane, experiments showed that most of these are closed when the neuron is at rest, making it very difficult for the sodium ion to get into the cell. However, a series of further experiments were carried out that disproved this theory: sodium ions were radioactively labelled, so that they could be marked and tracked, and lo and behold it appeared that eventually these ions had made their way inside the cell.

The answer to this puzzling question lay in the discovery of a very clever pump – the sodium-potassium pump – which actively ejects the sodium ions back out of the cell, rather like a disgruntled bouncer throwing someone out of a night club! Therefore, although these positive sodium ions can sneak in through some of the open channels, they quickly get expelled again. This ensures that uneven balance of positive and negative ions, thus maintaining the all-important negative resting potential. This pump requires a lot of energy to do this job, which explains why neurons are so hungry. Despite the fact that they account for only about 2 per cent of body weight, neurons use around 20 per cent of our energy output.

THE NEURON IN ACTION

So, we have established that the neuron has to work quite hard merely to remain in its resting state – but what happens when it actually fires? The mechanism is ingeniously simple, and again, rests on the groundbreaking work of Hodgkin and Huxley. They discovered that many of the sodium channels in the membrane are *voltage-gated* – meaning that they open and close in response to a change in voltage. Remember that when the neuron is at rest it has a negative voltage across it. This keeps the channels closed. However, when the neuron becomes excited by some kind of stimulus – for example, light or sound – there is a build-up of positive charge inside the top end of the axon. This causes the sodium channels in that part of the axon to unlock, thus allowing a sudden and massive influx of positive sodium ions into the cells, causing the inside to switch from negative to positive.

One thing that makes all this work so elegant is that the switch in voltage is a momentary phenomenon. The voltage-gated sodium channels snap shut again as soon as the voltage builds up and that section of the axon is quickly restored to its resting state, thanks to the actions of the pump and to the rearrangement of other ions. However, the real beauty of this mechanism is that the temporary switch in voltage triggers the next channels in line to open, which causes the same flip in voltage in the next section of the axon. This cycle of events continues all the way down the axon until it reaches the end (*see* Fig 2.3 opposite). This wave of electrical activity is described as a *neural impulse*, or *action potential* – and this is how all signals in the brain and across the whole body are carried.

Before we move on, there are two more very important things to know about how a nerve impulse travels. The first is that the signal can only ever move in one direction – from the cell body to the terminal buttons. This is because when the voltage-gated sodium channels snap shut, they stay locked for a brief time (a bit like the door on a washing machine at the end of its cycle). This ensures that the wave of electrical surge does not reverse back on itself. Having this control over nervous signals is vital, because otherwise there would be chaos in the brain. A second key point is that a neural impulse is an "all-or-nothing" action – it either goes or it does not go – and the strength stays exactly the same all the way down the axon. It is similar to the action of pressing a doorbell or taking a photograph, in that there are no half measures – it either happens or it does not. This essentially means that all brain activity – every thought we have and every move we make – is driven by a huge number of on-off switches.

HOW DO BRAIN CELLS COMMUNICATE WITH EACH OTHER?

So far, this chapter has explained what a neuron is and how it is able to carry the electrochemical signals that allow us to move, think and even just breathe. However, each neuron is only one tiny part of an extremely intricate jigsaw. Simply reading and interpreting the words on this page requires the finely tuned coordination of hundreds of thousands of neurons. To understand how this happens, we need to look next at how neurons communicate with each other.

Remember that a nerve impulse only ever travels in one direction, from the cell body, along the axon, to the terminal buttons. We can think of the nervous system as one huge network of one-way streets, with each individual street leading onto and off thousands of other streets. To give you an idea of how complex this system is, there are estimated to be something like 10^{14} of these junctions in the human brain.

Unlike the electrical circuitry in a television or computer, in which every wire is in some way physically connected to the next one, neurons in the brain never actually come into direct contact with each other. Instead, at each neuronal junction there is a very small gap called a synapse, across which the signal is transmitted. Some synapses are incredibly small, around 3.5 nanometres (nm) wide, which means it is possible for the nerve impulse to essentially jump from one neuron to the next. However, this type of synapse – the *electrical synapse* – is uncommon in the adult human brain and the vast majority of neurons communicate via *chemical synapses*.

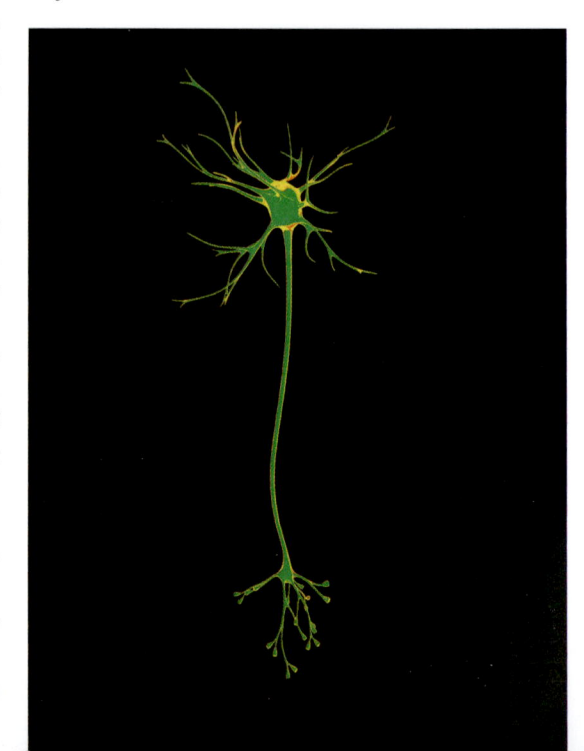

Chemical synapses are a little bigger (around 20–40 nm) and in this case the neuronal signal is relayed by means of special chemical messengers called *neurotransmitters*. The behaviour and qualities of these simple chemicals are absolutely fundamental to the complex functions of the human brain. Manipulating the actions of neurotransmitters, through recreational or medicinal drugs, or even naturally occurring toxins, can lead to anything from a subtle change in mood to coma or death. It is not surprising, then, that they have been such a focus of scientific study since the first neurotransmitter was discovered in 1913.

Neurotransmitters are stored in little membrane sacs, called vesicles, inside the terminal buttons of the neuron. To refer back to our earlier analogy, the terminal buttons are like the junction at the end of a one-way street. When the signal – the nerve impulse – arrives at this junction, it causes the neurotransmitters to be released into the fluid-filled synapse (*see* Fig 2.4, opposite). The neurotransmitter molecules then float across the synapse, where they meet the next neuron, and bind themselves to special proteins called receptors – much like a boat docking at port. Once this binding has occurred, the message from neuron A (the presynaptic neuron) has successfully reached neuron B (the postsynaptic neuron).

Below: A nerve impulse travels from the cell body, along the axon, to the terminal buttons.

Fig. 2.4: The chemical synapse

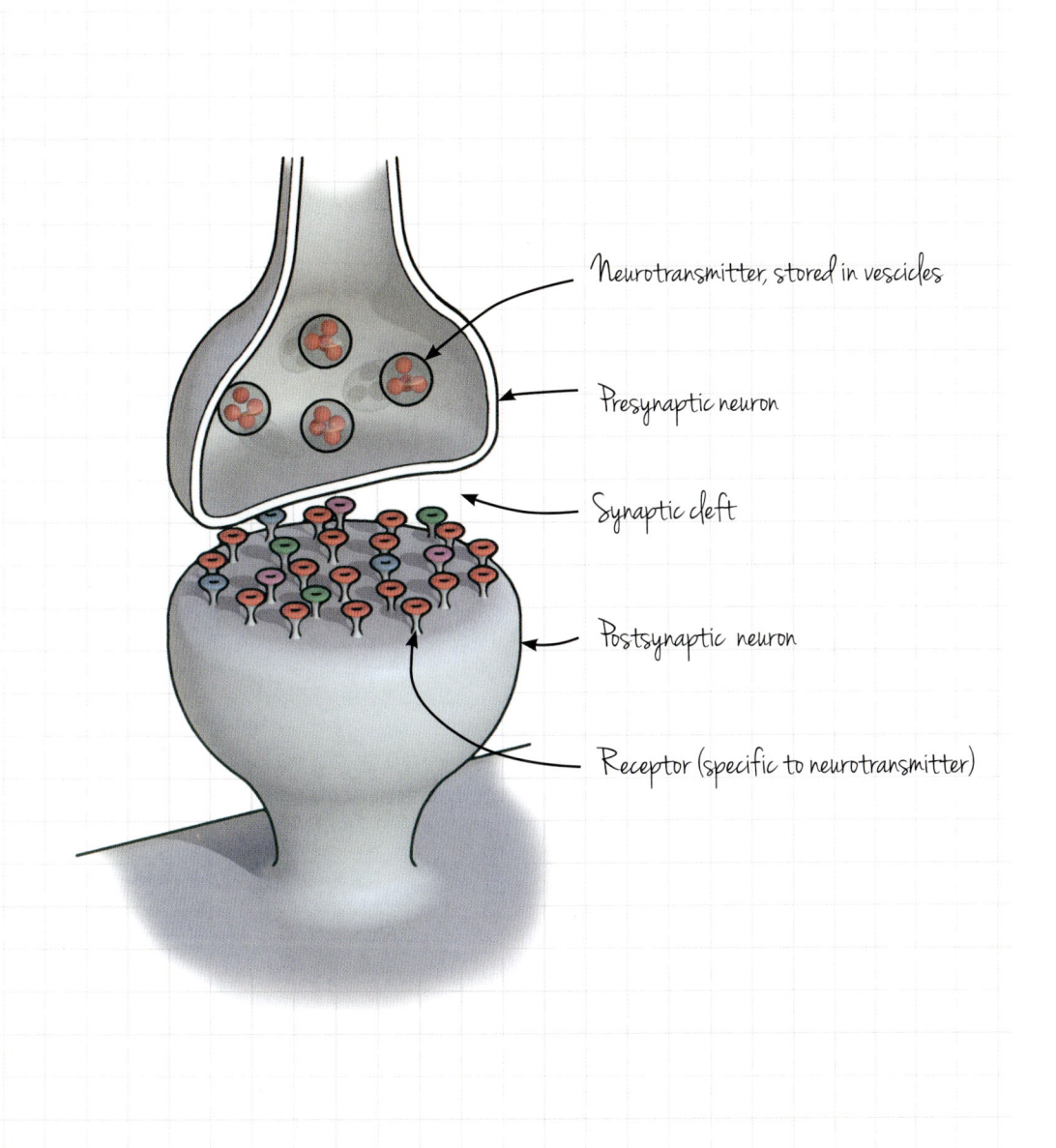

- Neurotransmitter, stored in vescicles
- Presynaptic neuron
- Synaptic cleft
- Postsynaptic neuron
- Receptor (specific to neurotransmitter)

How does the brain work?

Fig. 2.5: Synaptic communication

Action potential

Ca^{2+}

Ca^{2+}

Ca^{2+}

Ca^{2+}

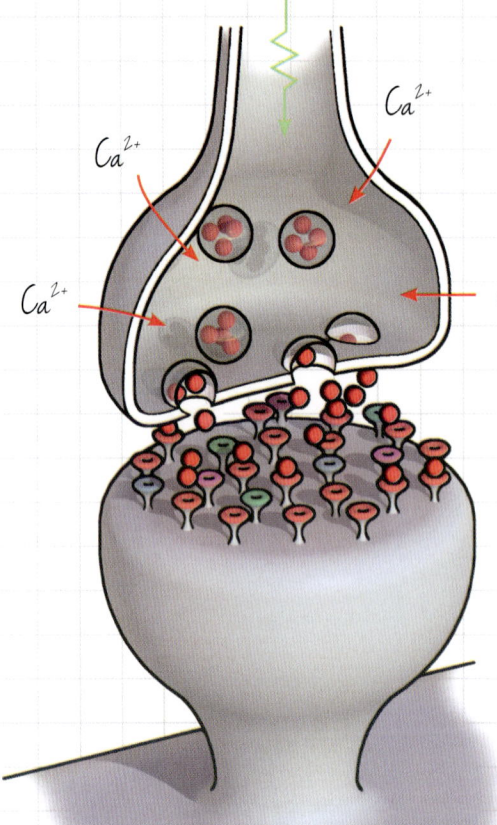

1: Arrival of action potential causes calcium gates to open and calcium flows in

2: Presence of calcium causes neurotransmitter to be released

3: Neurotransmitter binds with receptor causing excitation (EPSP) or inhibition (IPSP)

For most neurons in the brain, this will be just one of thousands of inputs, and the postsynaptic neuron must continually compute all of the incoming information to determine whether or not it fires itself. This neuron in turn will be just one of thousands of inputs onto the next neuron in the communication chain. When you consider the complexity of this system, it is extraordinarily impressive that it works so fluently and accurately.

WHAT ARE NEUROTRANSMITTERS AND WHAT DO THEY DO?

Central to this complex interaction between neurons are *neurotransmitters*, so let us look in more detail at the different types and what we know of their role in our daily lives. A neurotransmitter is defined as a chemical that (1) is made and stored inside a neuron, (2) is released on stimulation, and (3) will cause a response in the postsynaptic neuron. Nowadays, the names of the most common neurotransmitters – for example, dopamine, serotonin, noradrenaline – are so frequently cited in the media that they are almost becoming part of everyday vocabulary. However, at the turn of the twentieth century, almost nothing was known about these vital chemical messengers.

The first neurotransmitter to be formally identified was acetylcholine. Pioneering work in the early 1900s by both the English pharmacologist Sir Henry Dale and the German psychobiologist Otto Loewi established that this chemical provoked a muscle contraction when released by a motor neuron. Blocking the actions of acetylcholine – using nerve gases, for example – causes paralysis. We now know that this is the major chemical messenger for all voluntary muscle contractions, but it has also been found to regulate heart contractions and to play a very important role in memory: levels are significantly lower in people with Alzheimer's disease.

Noradrenaline was also an early discovery and, as its name might suggest, this neurotransmitter is involved in our body's emergency responses. It is released in times of stress and excitement, causing the heart to beat faster and the lungs to open up so that we can take in more oxygen. It also helps our brain to work faster, generally lifts our mood and makes us feel alert and focused. As we will see in the next chapter, drugs that increase levels of this neurotransmitter can improve our mood, lower our appetite and make us feel more alert.

"If there were a celebrity among brain chemicals, it would be dopamine," said the clinical psychologist Vaughan Bell, writing for the *Guardian* in 2013. Dopamine does indeed hit the headlines regularly, primarily because it is known to be important in addiction and pleasure-seeking. For example, the latest dopamine-related headline at the time of writing is that a dopamine surge may be responsible for over-eating. While it is true that dopamine is a key player in our brain's sense of pleasure (*see* Chapter 3 for more discussion around

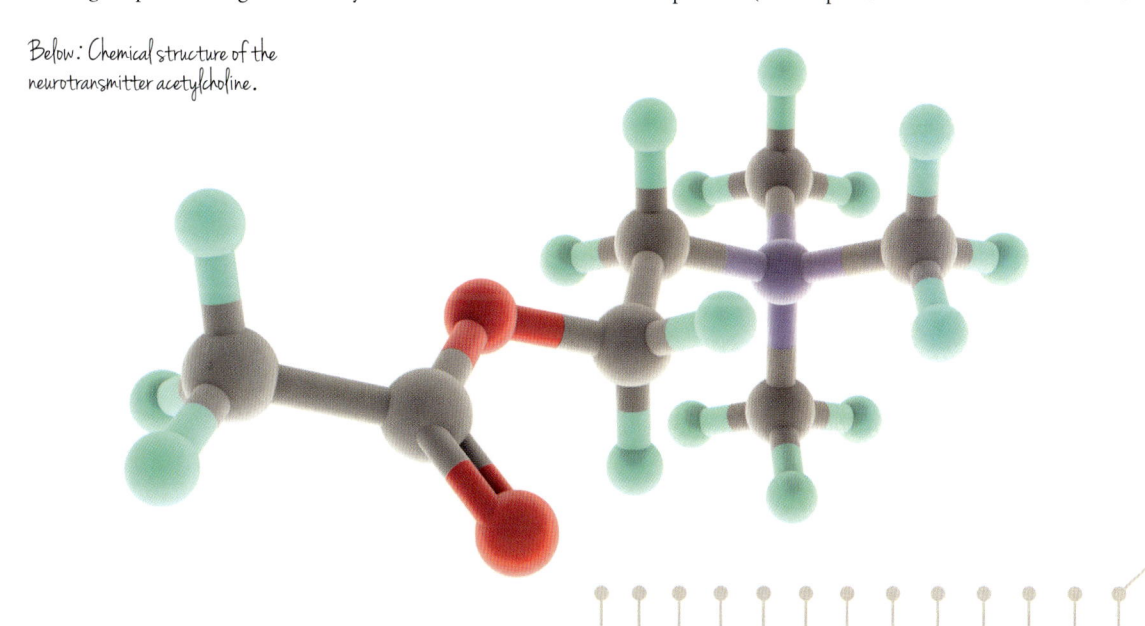

Below: Chemical structure of the neurotransmitter acetylcholine.

this), it also has other very important functions, including movement, attention and learning, and is known to be the major neurotransmitter disrupted in Parkinson's disease.

A particularly versatile neurotransmitter is serotonin, also referred to as 5-hydroxy-tryptamine (5HT). A large amount of serotonin is actually found outside the central nervous system in the gut, where it regulates the movements of the intestine. However, there are important serotonin pathways in the brain as well, and this chemical is thought to significantly influence functions ranging from sleep, appetite and temperature regulation through to mood, memory and perception. Among other things, serotonin has been linked with migraines, eating disorders and depression. One interesting fact about serotonin is that it is directly derived from tryptophan, a protein found in our diet – particularly in nuts and seeds, eggs and poultry. When levels of tryptophan are deliberately lowered in healthy volunteers, they can experience depression, poor memory and a higher tendency for aggression.

Most of the neurotransmitters discussed so far are primarily *excitatory* – that is, they tend to *stimulate* neuronal activity. Given how many hundreds of thousands of neurons and synapses there are, this could lead to complete mayhem if it was not for gamma amino butyric acid (GABA). This very important neurotransmitter is an *inhibitor* and its main job is to keep neural communication under control so that signals can be directed along particular pathways in the brain. If every signal that arrived at every neuron caused it to fire and there was no mechanism for turning it off, then everything would be firing all the time. Imagine if every time you turned a light on in your house it made all of the others come on, as well as the oven, the washing machine, the kettle, the radio and so on!

Indeed, when GABA function is disrupted in an individual, they are likely to suffer a seizure, or an epileptic fit, due to a temporary loss of control over the waves of activity in the brain. It is not surprising to learn that drugs such as alcohol and valium that boost the actions of GABA tend to increase the amount of neuronal inhibition. In other words, they switch off or reduce the number of nerve impulses that occur throughout the brain. This is why these substances can make people feel calm and sleepy, or in excessive doses can lead to coma or even death.

PATHWAYS THROUGH THE BRAIN

Thus, we have established that acetylcholine makes our muscles contract and helps us to form memories; serotonin influences our sleep, temperature and mood; and dopamine helps us to experience pleasure – all quite different jobs in the human brain and body. But how does each of these neurotransmitters manage to have such a defined role within the brain when they are all simply acting as a messenger between neurons? One important factor is that when a neurotransmitter is released into the synaptic cleft, it can only bind with a receptor that has been designed to accept it. So GABA will only bind with a GABA receptor and noradrenaline will only bind with a noradrenaline receptor. This is called the lock-and-key hypothesis (*see* image opposite) and, as well as explaining how neurotransmitters have specific effects, it is also at the heart of how drugs can influence the brain.

The other reason for such specificity is that each neurotransmitter works within its own system. If we liken the brain to a big city with many distinct areas, then each of the neurotransmitter systems can be compared to specified train, bus or underground routes. So, in London, England, for example, the Piccadilly Line will take a passenger from Heathrow to Piccadilly Circus or Kings Cross – but if someone

Below: Serotonin is derived from tryptophan, a protein found in nuts and seeds.

wanted to go from St Paul's Cathedral to Marble Arch, they would need to take the Central Line. Similarly, there are a number of dopamine pathways in the brain – for example, one that goes from the mid-brain to the frontal lobes – and each is a precisely defined route of connected neurons that only uses dopamine as its messenger. We know of four dopamine pathways in the brain and each regulates a particular behaviour – so, for example, one is associated with pleasure and reward and another with the control of movement. Similar principles are true of other neurotransmitters.

In this chapter we have considered the different cells in the brain, how they store and transmit an electrical charge, how they communicate with one another and, finally, the selective way in which different chemical messengers mediate particular thoughts and behaviours. The next chapter will look at what happens when these neurotransmitters are disrupted or deliberately manipulated and how this can change the speed, nature and content of our thought processes, as well as other bodily functions.

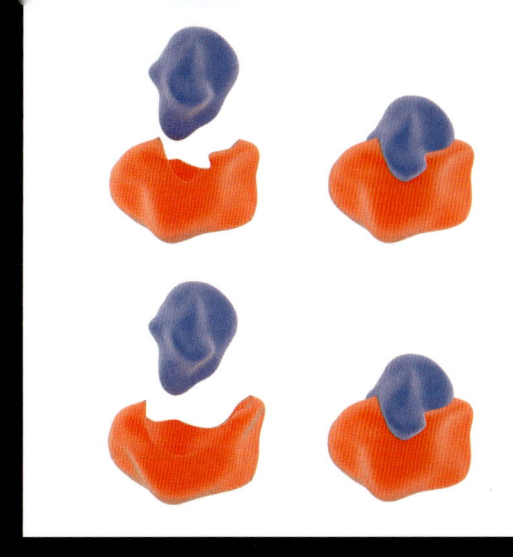

Above: Neurotransmitters are believed to bind to receptors via a lock-and-key mechanism.

MYTH BUSTER: WE USE ONLY 10 PER CENT OF OUR BRAINS

Most of us will have at some point heard the claim that we use only 10 per cent of our brains. This seems like a ridiculous suggestion and yet the 10 per cent myth has stubbornly refused to die and continues to appear in media, adverts and the commercial secrets of successful businesses. Maybe the appeal of this myth is that we all want to believe that we have secret superhuman capabilities. Indeed, this idea is the central premise of the 2014 film *Lucy*, in which the protagonist accidentally unlocks the other 90 per cent of her brainpower and is able to perform extraordinary cognitive feats.

However, it is not just the entertainment industry that perpetuates this myth: Sophie Scott, neuroscientist at University College London, was once told at a first aid training session that most head injuries are not too serious because we use only 10 per cent of our brains anyway. This is clearly not the case, as anyone who has experienced, or seen, the effects of brain injury, dementia or strokes will be able to attest. Damage to almost any area of the brain will lead to a loss of function and sometimes even a very small amount of damage can cause devastating effects. So, clearly we use most, if not all of the brain that we have.

Brain scans also help to debunk this myth. We now know that simply tapping a finger or listening to a piece of music causes activation in far more than 10 per cent of the brain. Ironically, one of the most fascinating pieces of research has shown that *savant* skills, the closest to superhuman abilities we see in the real world, can be unlocked by temporarily *closing off* areas of the brain (albeit with other negative consequences). So, if anything, using less of our brain might be what we actually need in order to perform extraordinary feats.

3. HOW AND WHY DO DRUGS AFFECT THE BRAIN?

ARE YOU SOMEONE WHO NEEDS A MORNING CUP OF COFFEE TO HELP YOU WAKE UP? THIS IS JUST ONE SIMPLE WAY TO CHANGE YOUR BRAIN CHEMISTRY.

In March 2015, city officials closed down a restaurant in Wakayama City, Japan, after five diners suffered with vomiting and breathing difficulties. Each had chosen to eat *fugu*, the Japanese puffer fish, which is considered by many to be a delicacy, a single serving costing hundreds of dollars. Part of the appeal is that it can cause a tingling of the lips, but when it is not prepared properly it is highly toxic. In fact some parts of the puffer fish contain a poison – a *neurotoxin* – that is powerful enough to kill an adult. Consequently, a sushi chef has to train for two years and receive a special licence in order to serve it. The potent neurotoxin present in puffer fish is called *tetrodotoxin* (TTX), and its dramatic effects are caused by the fact that it blocks the sodium channels in neurons, thus halting all communication in the affected nerves. A small amount can lead to the lip tingling that many find quite pleasant – but too much causes paralysis and possibly death.

CHANGING BRAIN CHEMISTRY

Japanese puffer fish are not alone in being able to play with the chemistry of the nervous system. Botulinum toxin (BTX) is produced by a bacterium called *Clostridium botulinum*, which can occur in contaminated food or in wound infections, and which causes the potentially fatal disease, botulism. This particular neurotoxin prevents the neurotransmitter, acetylcholine (*see* page 33), from being released. This in turn means that there is no way for the neurons to communicate with muscles, which leads to paralysis. However, this potentially lethal toxin is now produced commercially for use in medical and cosmetic settings. As a medicine, it can be helpful in treating particular muscular conditions and has been found to relieve chronic migraines. In the cosmetic field, it is used for Botox treatment, in which injecting a small amount into facial muscles prevents contraction, leaving the face looking smoother and younger. Unfortunately, this treatment also means that the face is less able to be expressive, so smiles and frowns may appear a little half-hearted.

For many centuries, humans have known how to take advantage of naturally occurring substances that alter brain chemistry. Christopher Columbus reportedly discovered that Native Americans living in South America were able to make poisonous arrow tips using a substance named *curare*, which was extracted from certain species of woody vines. Our ancestors were also quick to discover that there were other, more pleasant effects of particular substances. There is evidence that opium poppies, which produce a natural morphine, have been cultivated since Neolithic times, and this substance was certainly being used recreationally in China by the fifteenth century. The popular English pint of beer has an even longer history: the production of fermented drinks is said to date back to at least 7,000 BC!

Nowadays, of course, we have the knowledge and facilities we need to artificially synthesize all these naturally found chemicals and many more. The number of mind-altering drugs is said to be increasing faster than

Fig. 3.1: Sites of drug action

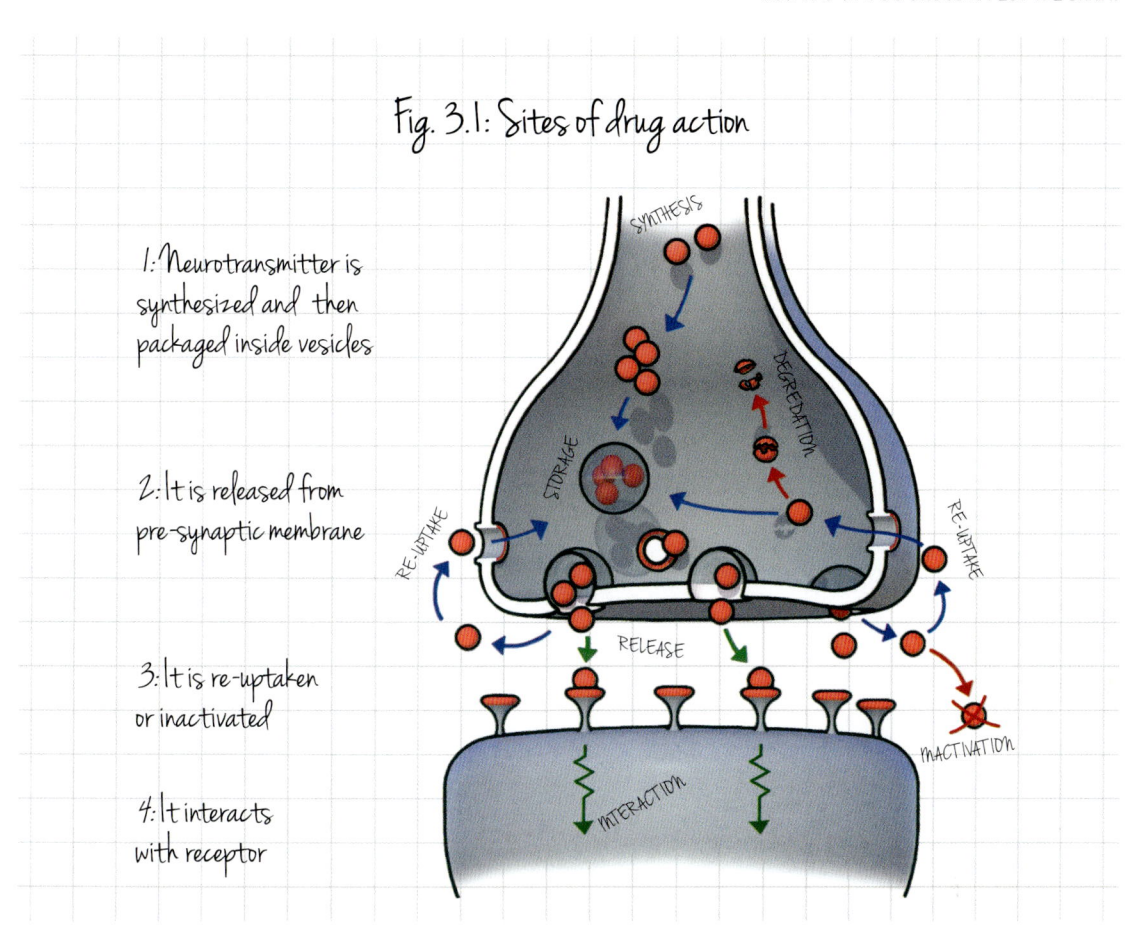

1: Neurotransmitter is synthesized and then packaged inside vesicles

2: It is released from pre-synaptic membrane

3: It is re-uptaken or inactivated

4: It interacts with receptor

---

it is possible to make them illegal, and the pharmaceutical companies that research and produce drugs for medical use are unlikely to be going out of business any time soon. Why is the production of psychoactive drugs such a lucrative industry? And can drugs effectively treat people with mental health conditions or degenerative disease? This chapter explains the basic principles of *neuropharmacology* – how drugs are able to change the chemistry of the brain for better or worse.

DRUGS FOR NEUROLOGICAL DISORDERS

What each of these natural and synthetic substances has in common is that they alter normal communication in the nervous system, usually by interfering with the neurotransmitters at the synapses. Remember that in the normal cycle of events: (1) a neurotransmitter is synthesized and stored in the pre-synaptic neuron; (2) when the neuron is stimulated, the neurotransmitter is released into the synaptic cleft; (3) the neurotransmitter briefly binds with the receptor, which causes a response in the post-synaptic neuron; (4) after the neurotransmitter has done its job, it is destroyed or taken back up into the cell, where it can either be broken down or re-used. Any chemical agent that disrupts these processes – from botulinum to alcohol – will change the natural course of events, thus altering the mind and/or behaviour. Some substances promote or increase activity in the synapse and these are called *agonists*; others reduce or block activity – these are called *antagonists*.

Let us consider a really impressive example of how modern medicine has manipulated the first of these stages, when the neurotransmitter is synthesized and packaged up ready for use. Parkinson's disease is a neurodegenerative condition that has a profoundly debilitating effect on movement and later impacts on mood and cognition. Around 1960, the Austrian biochemist Oleh Hornykiewicz showed that people with Parkinson's disease had extremely low levels of dopamine due to loss of cells in the *substantia nigra*, an area of the brain that plays an important role in initiating and controlling movement. During the early stages of the disease, a sufferer will experience mild tremors, have difficulties initiating movements and will usually walk with a characteristic slow shuffle. However, as the disease progresses, the symptoms become much more severe – wild uncontrollable shaking, severe difficulties speaking, a fixed mask-like facial expression and extreme stiffness in the limbs. Michael J. Fox, an American actor who was diagnosed with Parkinson's disease when he was just 30, explains the effects: *It makes me squirm and it makes my pants ride up so my socks are showing and my shoes fall off and I can't get the food up to my mouth when I want to.*

Given the low levels of dopamine, an obvious solution might be to simply administer more of this neurotransmitter to people with the condition. However, studies showed that this would not work, because dopamine cannot cross the blood brain barrier – a special layer of protection that

Below: Muhammed Ali and Michael J. Fox have both been diagnosed with Parkinson's disease.

prevents many substances from passing from the blood into the brain. Hornykiewicz, along with another scientist, Greek-American George Cotzias, took an ingenious alternative approach. They found that L-dopa, the primary building block for dopamine, could cross the blood brain barrier and that when it did, it stimulated the synthesis of more dopamine. The effect on patients was quite dramatic – suddenly they were able to move and speak again! Although there are problems with longer-term use of L-dopa, it is still considered something of a wonder drug for those who suffer with this devastating condition. For someone like Michael J. Fox, this medication is the difference between him being able to function and not function. So, L-dopa remains the principal treatment for Parkinson's disease and continues to be used alongside newer surgical techniques.

Alzheimer's disease is another neurodegenerative condition in which cell loss leads to a significant drop-off in levels of a neurotransmitter, with catastrophic consequences. In this case there is progressive destruction of the *basal nucleus*, which produces acetylcholine, the neurotransmitter that enables us to lay down new memories. Once again, we cannot simply replace the missing acetylcholine, because it will not cross the blood brain barrier; yet attempts to boost production by administering the precursor, choline, have also been relatively ineffective.

So, in Alzheimer's disease scientists have been forced to take a different approach: instead of trying to increase levels of acetylcholine, they have found a way of maximizing what is naturally produced. Under normal conditions, an enzyme called *acetylcholinesterase* breaks down the acetylcholine soon after it has been released into the synaptic cleft (to ensure that it binds to receptors for only a short time). Drugs that stop this enzyme from working are called *acetylcholinesterase inhibitors*, or more simply AChEIs; these prevent the breakdown of acetylcholine, which means that more is available. Although there are often side effects with this drug, it can be very effective in reducing memory loss in some people and it is therefore widely used.

Of course, in both Parkinson's disease and Alzheimer's disease – and indeed other neurodegenerative conditions – the types of pharmaceutical treatments described here are only really a sticking plaster. They treat the symptoms – with

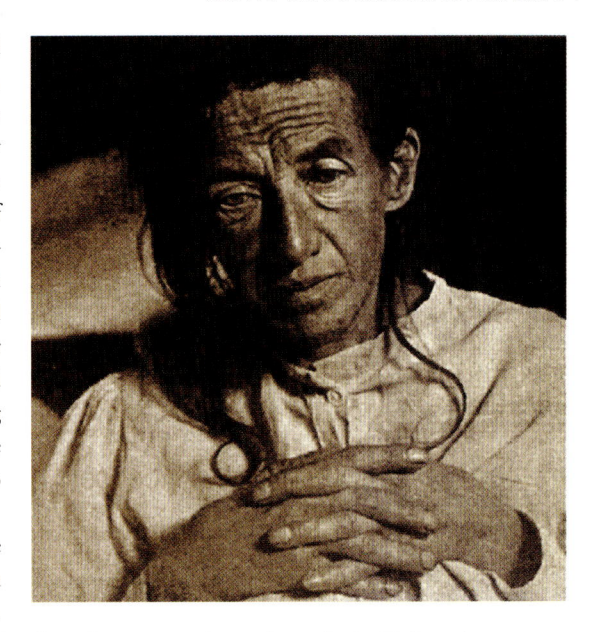

Above: German woman Auguste Deter (1850–1906) was the first person to be diagnosed with Alzheimer's disease.

varying degrees of effectiveness – by boosting levels of the missing neurotransmitters, but they do not get to the root cause. Consequently, there is a huge amount of ongoing research that seeks to understand and prevent or repair the underlying cell loss, so that the symptoms never take hold. Developments are promising, but in the meantime these strategies continue to be very helpful to many people.

DRUGS IN MENTAL HEALTH

While L-Dopa and AChEIs are not without their problems, the case for using medicines in neurodegenerative conditions is relatively secure: there is an observable drop in the levels of specific neurotransmitters and a clear effective strategy for replenishing supplies. But what are the arguments for taking a similar pharmaceutical approach to treating mental health conditions, such as depression, anxiety and schizophrenia? Although drugs are commonly used and are often very effective in psychiatric treatment, the rationale is less straightforward. A condition such as Parkinson's disease has a definitive and predictable set of symptoms and is relatively easy to diagnose. There is also strong and consistent evidence that dopamine levels are lower, as well as a clear mechanism for why this is the case. Neither of these things is true for any psychiatric diagnosis: the aetiology and symptoms vary hugely from person to person, the diagnostic criteria are less clearly defined and, most importantly, differences in brain chemistry are difficult to identify and show many inconsistencies.

That is not to say that drug treatments do not have an important place in mental health – there is a lot of evidence that for some individuals they are highly effective in alleviating distressing, debilitating and potentially life-threatening symptoms. And while some professionals suggest that psychiatric medication is prescribed too freely and is not always justified by the research, others have argued that the controversy and stigma attached to taking medications can have life-threatening consequences. For example, recent work in the UK by Damien Ridge and colleagues has shown that use of antidepressants often carries a huge stigma and is viewed as a "dirty little habit". Ridge argues that drug treatment can be the difference between life and death for some people. Given that suicide is currently responsible for more than 6,000 deaths every year in Britain and the Republic of Ireland alone, it is clearly important to carefully balance the risks against the benefits.

HOW ANTIDEPRESSANTS WORK

So, how do antidepressants actually change the brain chemistry? There are a number of different drugs that all work in a slightly different way, but each of them ultimately increases the activity of neurotransmitters that are important for mood – dopamine, noradrenaline and serotonin. Monoamine Oxidase Inhibitors (MAOIs), such as *selegiline*, were first used in the mid-1950s and act by preventing the natural breakdown of dopamine, noradrenaline and serotonin, thus increasing availability of these neurotransmitters in the synapse (stage 4 in the cycle described earlier, *see* page 37). They have been problematic in the past, in that they interact badly with some foods, causing nasty and potentially lethal side effects; however, newer versions appear to be much safer.

Tricyclic antidepressants (TCAs) were also very early effective antidepressant drugs, first coming to the market in 1957. This category of drugs also increases the activity of mood neurotransmitters, specifically serotonin and noradrenaline, but using a slightly different mechanism to the MAOIs. Instead of preventing them from being broken down, TCAs block the pumps that would normally transport them back into the cell, leaving higher levels of these neurotransmitters swimming around in the synapse. TCAs and MAOIs have now been mostly replaced by second-generation antidepressants, such as *fluoexetine* and *venlafaxine*, which are generally safer and better tolerated. Fluoexetine, which goes under the well-known brand name Prozac, is a Selective Serotonin Re-uptake Inhibitor (SSRI) and does exactly what its rather long-winded name suggests: it selectively inhibits the re-uptake of serotonin. In other words, much like the TCAs, it increases the levels of available serotonin by stopping it from being taken back up into the cell.

One of the controversies around antidepressants is that the symptomatic relief of most of these drugs appears to take the relatively long period of two to three weeks to come into effect. However, given that these substances work by making immediate adjustments to the neurotransmitter levels in the synapse, we might expect their effects to be fairly instant, much like those of drinking a glass of wine or smoking a cigarette. There are some recent findings that might explain this (*see* MYTH BUSTER box, page 51), but overall this state of affairs illustrates that overcoming depression is not simply a matter of topping up neurotransmitters.

CHLORPROMAZINE, BENZODIAZEPINES AND BETA-BLOCKERS

Just before the first-generation anti-depressants came onto the market, there was another big discovery – one that is often cited as the real birth of effective psychiatric drug treatment. *Chlorpromazine* was synthesized by Laboratoires Rhône-Poulenc, the French pharmaceutical company, as part of a long-term development programme for new antihistamines. It seemed to be such an effective sedative for people about to undergo surgery that a group of French doctors agreed to try it on a 24-year-old male psychiatric patient who was severely agitated and psychotic. The calming effect was dramatic and

HOW AND WHY DO DRUGS AFFECT THE BRAIN? **41**

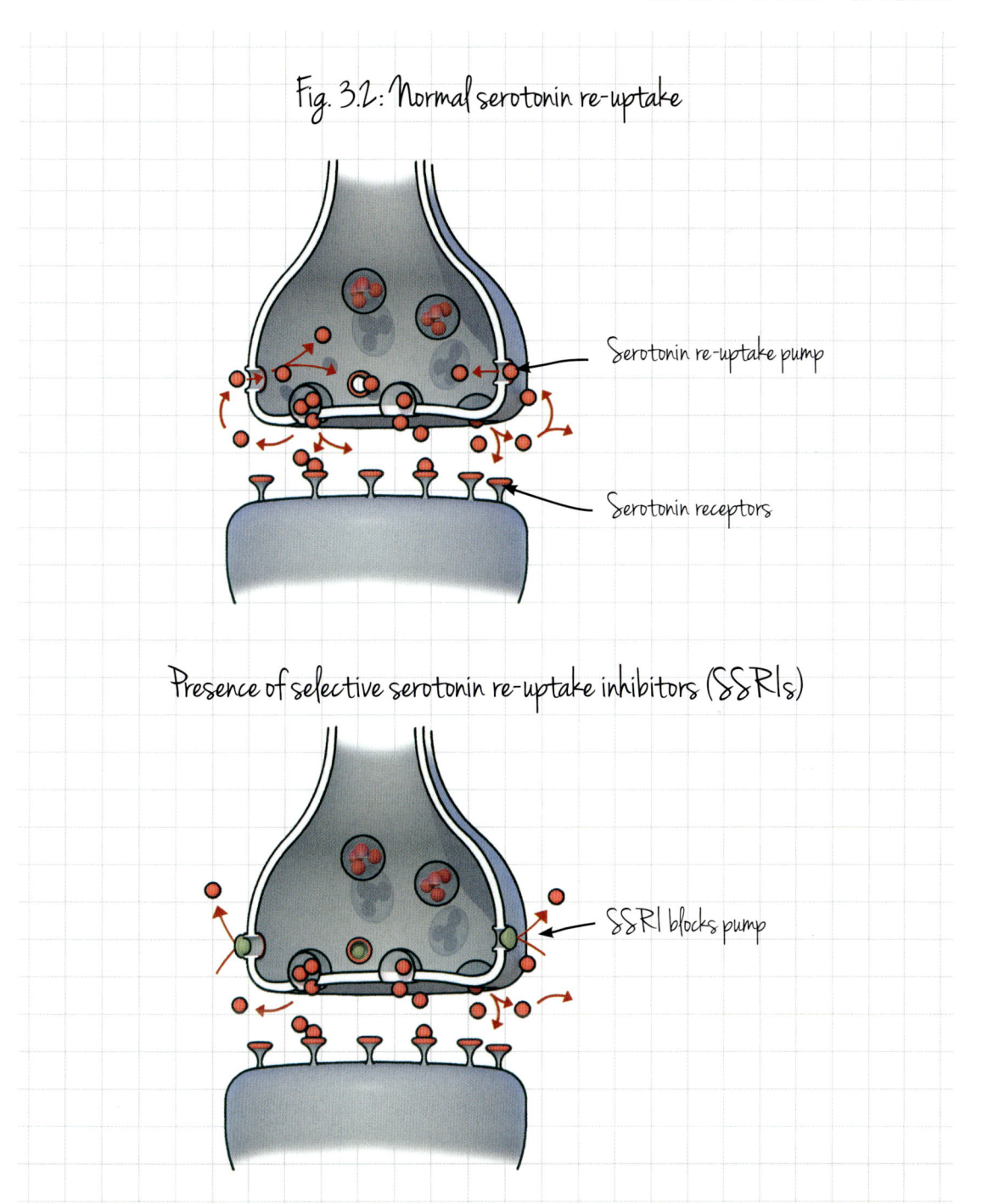

Fig. 3.2: Normal serotonin re-uptake

- Serotonin re-uptake pump
- Serotonin receptors

Presence of selective serotonin re-uptake inhibitors (SSRIs)

- SSRI blocks pump

immediate (although short-lived, so regular doses were required in order to maintain the effect). Very quickly the drug became marketed as an anti-psychotic cure and soon became the staple treatment for many patients with schizophrenia. Chlorpromazine, which went under the proprietary name of *Largactil* (meaning "large in action"), was seen as transforming the psychiatric wards into more peaceful and manageable places. However, as is the case with a lot of drugs, there were a number of unpleasant side effects to be contended with. Consequently, there has been huge investment into the development of newer and better *antipsychotic* drugs, and now there are many more for doctors and patients to choose from.

What is interesting about chlorpromazine is that it became popular before anyone really understood what it was doing or why it seemed to reduce psychotic symptoms. Some clues arose from the fact that heavy use of the drug caused some patients to experience unusual movements, reminiscent of those seen in Parkinson's disease, which raised some suggestions that the dopamine systems might be affected. Further experiments demonstrated that when chlorpromazine enters the blood stream it is able to get into the synapses, where it binds with and blocks dopamine receptors. So if it is possible to reduce many of the outward symptoms of schizophrenia by preventing the actions of dopamine, a logical conclusion might seem to be that schizophrenia occurs when there is too much dopamine in the brain. The *dopamine hypothesis* was big news for a while, but it became clear that this explanation was in fact far too simplistic. Many people now feel that although antipsychotics are a very useful treatment for some, they do not necessarily tell us much about what underlies the devastating condition that is schizophrenia.

Medication has also been very popular in the treatment of anxiety disorders – phobias, panic attacks, obsessive compulsive disorder (OCD) and so on – which lead to a host of very physical sensations, including a racing heart, sweating, tense muscles, nausea, as well as overwhelming feelings of fear and dread. Although anxiety often responds very well to psychotherapies, such as Cognitive Behavioural Therapy (CBT), there are also two drugs that are particularly helpful in alleviating symptoms – benzodiazepines and beta-blockers. The second of these – beta-blockers – act by blocking the receptors of noradrenaline, the neurotransmitter that normally stimulates the "fight-or-flight" response. By preventing noradrenaline from binding with the receptors, this prevents the normal physical symptoms of fear, so it is often used for acute anxiety conditions, such as stage fright (although it is also used very effectively in cardiovascular disorder, in which it slows the heart and reduces blood pressure).

Benzodiazapines, such as *diazepam* or *valium*, are a class of tranquilizers that work by enhancing the effects of GABA, the neurotransmitter that normally controls and dampens excitation in the nervous system. When they first came on the market in the mid-late 1960s, benzodiazepines were very popular, because they were seen to be mild, effective and safe. These little pills were famously described as "mother's little helper" by the Rolling Stones, and by the late 1970s they were the most commonly prescribed medication in the world: it is estimated that one in five women and one in ten

Above: Drugs can be a powerful contribution to the treatment of depression and anxiety.

men were taking them at this time. Benzodiazepines largely replaced barbiturates, which were a very effective sedative but highly addictive and potentially lethal, especially when mixed with alcohol; many believe that it was an overdose of barbiturates which caused the death of the American actress Marilyn Monroe in 1962. However, it turned out that, although benzodiazepines were safer – with a larger margin between the effective dose and the lethal dose – people very quickly became dependent on them, needing more and more of the drug simply in order to be able to function normally.

An outcome such as this is obviously a concern, but as always in these cases, the disadvantages must be weighed up against the potential benefits. Benzodiazepines are still viewed as a very useful short-term treatment for anxiety; they also make a very good pre-operative sedative, are an effective muscle relaxant and can prevent or reduce seizures. Current guidelines suggest that these drugs should be prescribed for a maximum of four weeks and only in cases in which anxiety is disabling and severe. Unfortunately, this directive is not always followed, leaving many people so dependent on the drug that without it they are susceptible to seizures and potentially even death. For this reason, it is vital that doses are reduced slowly and gradually and with good medical supervision.

ADDICTION, DEPENDENCE, TOLERANCE AND WITHDRAWAL

Many of the drugs we have discussed so far in this chapter – L-dopa, chlorpromazine, benzodiazepines – cause changes in the brain that lead to tolerance and dependence. This clearly has an impact on how they are used by the medical profession, but the same principles apply to recreational drugs, such as heroin and cocaine, where there is generally no medical control and potentially more serious consequences. The catalogue of famous drug-related deaths across the decades – from the guitarist Jimi Hendrix to Who drummer Keith Moon and the singer Amy Winehouse – provides a sad illustration of the dangers associated with substance misuse. Fatal overdoses are often the result of unpredictable changes in what the brain and body can tolerate. So, before we go on to look at how

Above: The withdrawal symptoms from nicotine can make it extremely difficult to quit smoking.

recreational drugs affect the synapses, let us first consider what is actually meant by the terms *addiction*, *tolerance* and *dependence*.

Addiction is considered to be a rather woolly term as far as neuroscientists are concerned. The word is generally used to describe the psychological state whereby someone feels increasingly compelled to carry out a particular behaviour, whether that might be taking drugs, gambling, shopping or any other activity. There is sometimes a fine line between regular pursuits that provide intense enjoyment and those behaviours to which someone might be addicted. Love and music have both been described as having addictive qualities, and indeed trigger many of the same brain areas that recreational drugs do, but few people would end up in therapy for spending too much time listening to the

HOW AND WHY DO DRUGS AFFECT THE BRAIN?

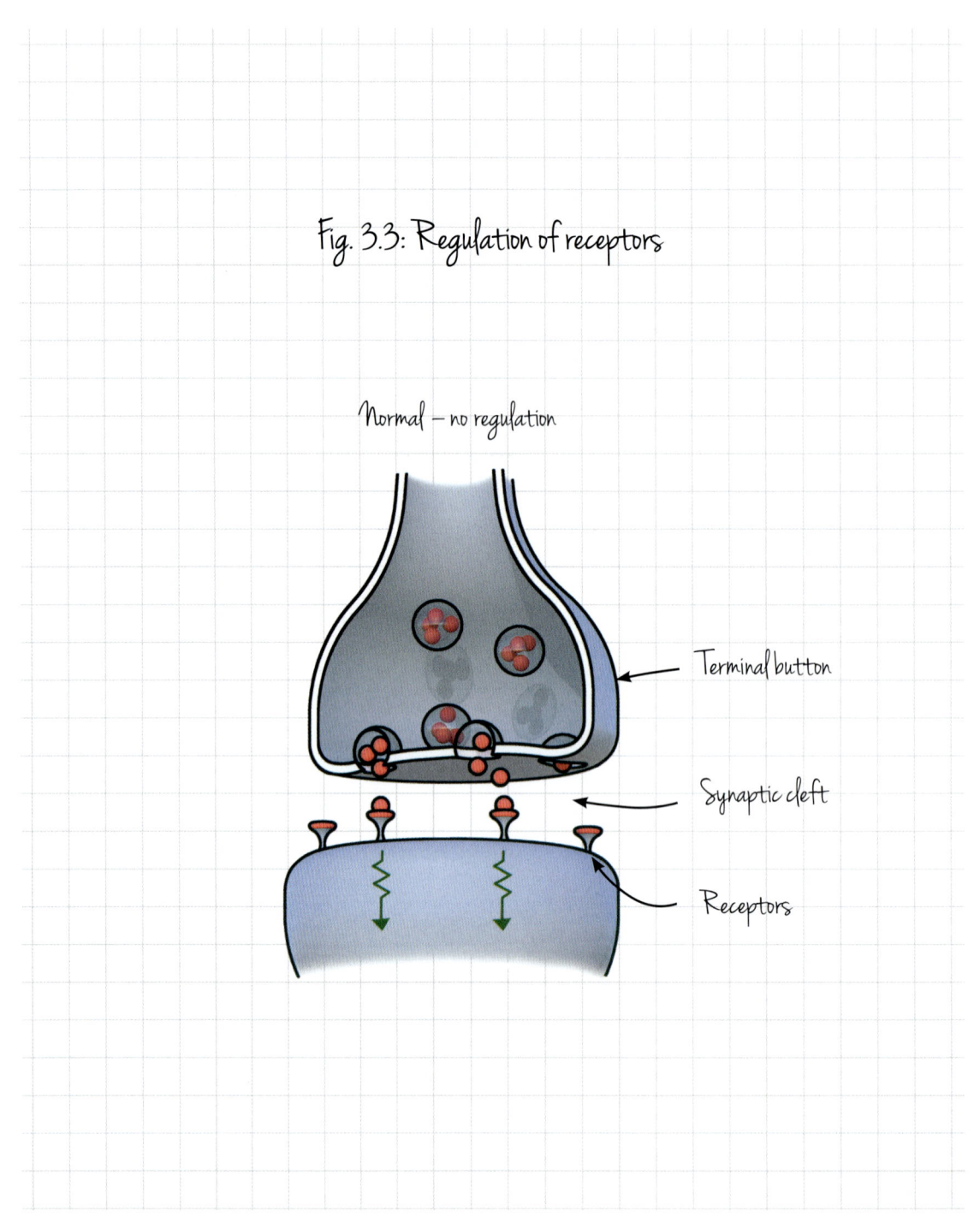

Fig. 3.3: Regulation of receptors

Normal — no regulation

- Terminal button
- Synaptic cleft
- Receptors

Fig. 3.4: Presence of agonists

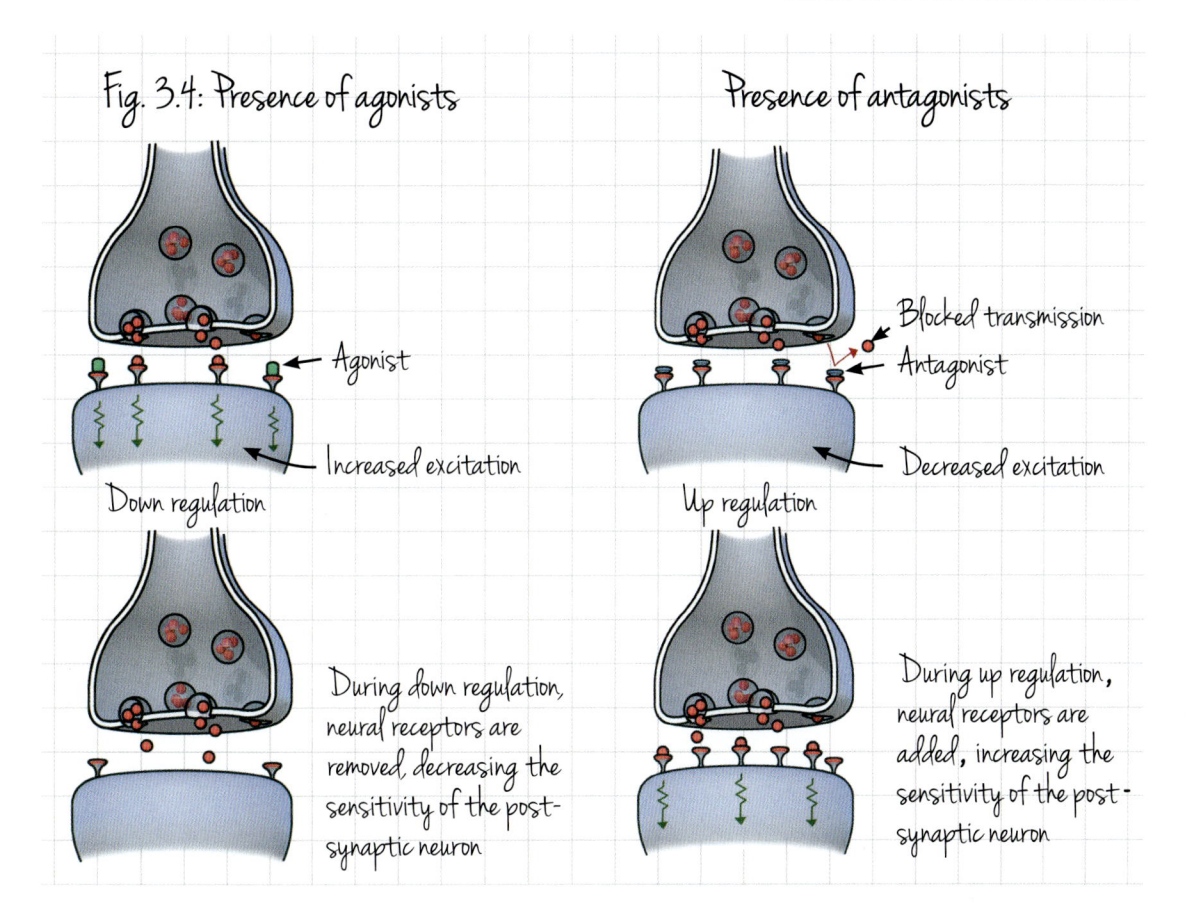

← Agonist

Increased excitation

Down regulation

During down regulation, neural receptors are removed, decreasing the sensitivity of the post-synaptic neuron

Presence of antagonists

Blocked transmission
← Antagonist

Decreased excitation

Up regulation

During up regulation, neural receptors are added, increasing the sensitivity of the post-synaptic neuron

---

Beatles! Thus, the term addiction is generally used to refer to a situation in which a person's life becomes so overwhelmed by pursuing a particular activity that it interferes with their life, and causes distress to themselves and their families.

Tolerance and dependence are much easier to pin down scientifically, because they describe a situation in which the brain physically adapts to the continued presence of a particular substance. Take L-dopa as an example: we saw earlier that this increases the amount of dopamine in the synapses, so the brain's adaptive response to this is to shut down or remove some of the receptors. This process, called *down-regulation* of receptors, means that the synapses gradually become less responsive to the dopamine and the symptoms return. A higher dose of the drug is then required to relieve the symptoms, but the process of down-regulation happens again, and so it goes on. Sometimes the situation is the opposite – for example, the antipsychotics we discussed earlier cause dopamine receptors to be blocked so that dopamine has less effect. The brain adapts to this by increasing the numbers of receptors – *up-regulation* – so that the synapses re-set themselves to their un-medicated levels of sensitivity, which means that the symptoms return.

This adaptation process in synapses happens with many drugs – antipsychotics, benzodiazepines, heroin, nicotine, alcohol, caffeine, cocaine and so on. As well as leading to tolerance, so that increased doses are needed in order to achieve the same effects – whether that be a "high" or symptom relief – this process also means that removing the drug completely will cause significant withdrawal symptoms. The change in numbers of receptors due to up- or down-regulation means

46 HOW AND WHY DO DRUGS AFFECT THE BRAIN?

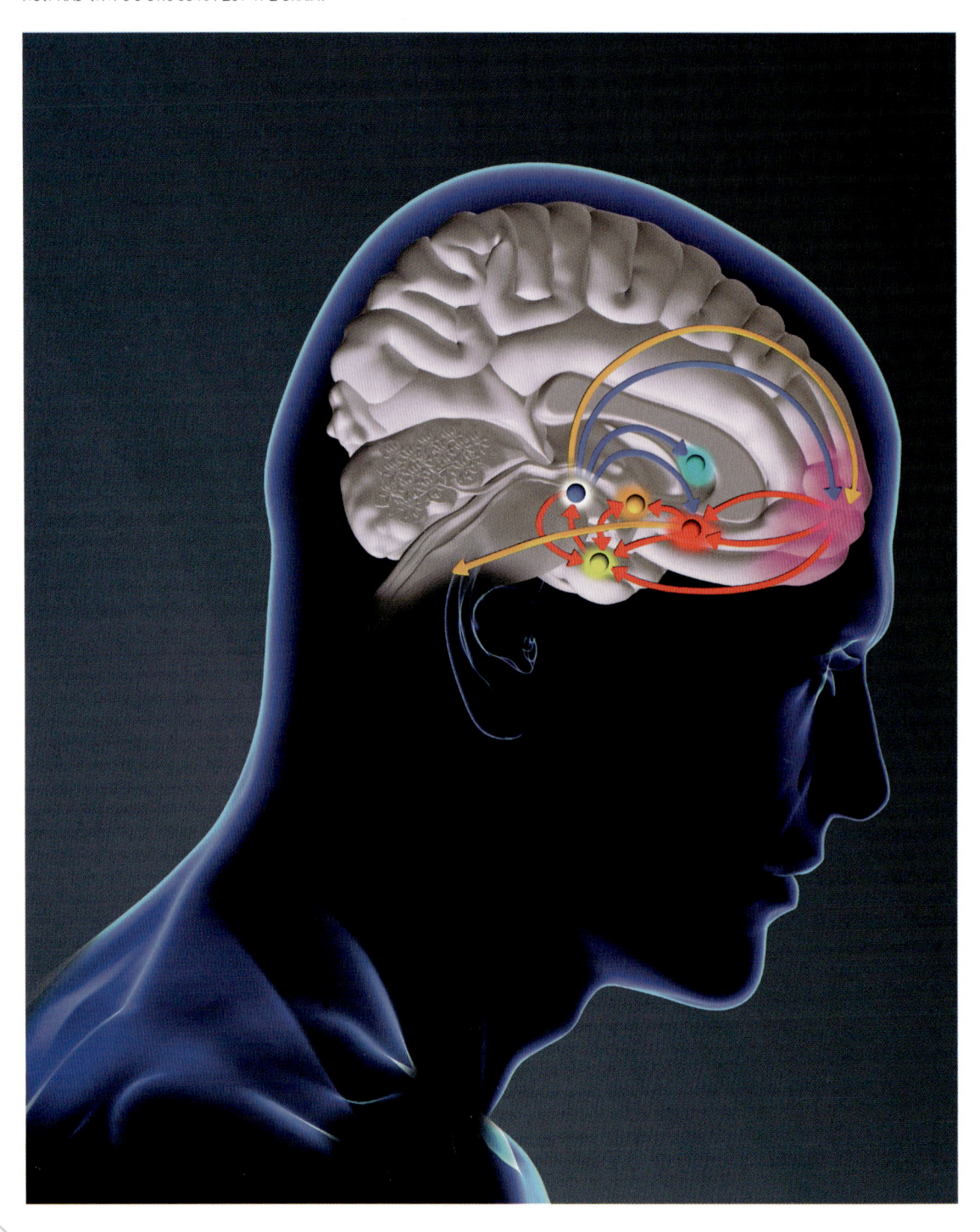

that when the drug is not present the synapses are hugely under- or over-activated. However, this is a reversible process, so although there will initially be withdrawal symptoms when the drug is removed, these do wear off, as the brain re-moulds itself back to its un-medicated state.

RECREATIONAL DRUGS

It has been estimated that nearly one in three British adults have taken an illegal mind-altering substance. Far more have drunk alcohol, smoked a cigarette or had a cup of tea or coffee. So what is it about these substances that make them so inviting, especially when they can be quite costly, both in terms of health and wealth? The answer lies in an intriguing network in the brain, which has been dubbed the *reward pathway*. This is a bundle of dopamine-containing neurons that run from the ventral tegmental area, a region deep in the brain, to the nucleus accumbens and then on to the prefrontal cortex. Brilliant work by a Canadian surgeon named Wilder Penfield (*see* page 13) in the mid-twentieth century showed that if you electrically stimulated this pathway in a conscious human, they would giggle and blush and describe an intense though non-specific pleasurable sensation. When these same stimulators were fitted into rats' brains, it was found that the rats would press the levers that triggered these electrodes up to 6,000 times an hour with no sign of letting up!

We now know that this reward pathway is activated in any pleasurable activity, be it eating and drinking, playing computer games, making love or listening to music. Also, it may come as no surprise to find that every type of

Opposite: Psychoactive drugs hijack the brain's natural reward system.

Above: The reward pathway is activated when we do anything enjoyable, such as eating chocolate or playing a game.

Above: Ecstasy acts as both a hallucinogenic and a stimulant.

recreational drug has a hotline to this reward system, which is why these substances are capable of inducing pleasurable feelings, ranging from a gentle warm buzz to a sudden and intense high. Each drug affects a range of neurotransmitters, which means that it has its own unique and characteristic effect on the mind; we will briefly consider these below. However, as part of their effect, they all in some way increase dopamine within this reward pathway, which is why they can be so compelling.

ILLEGAL HIGHS

We will look at the mind-altering effects of illegal drugs in more detail in Chapter 14, but let us briefly consider how the different classes of drugs affect the neurotransmitter systems. Stimulants, which include caffeine, nicotine, amphetamine and cocaine, all work by increasing the availability of dopamine and the fight-or-flight neurotransmitter, noradrenaline. For example, amphetamine stimulates their release and cocaine prevents them from being taken back up into the cell. Conversely, depressants, or sedatives, slow the central nervous system down, usually by enhancing the actions of the inhibitory neurotransmitter GABA. This class of drugs includes alcohol, tranquilizers (such as benzodiazapines) and the date-rape drug, gamma-Hydroxybutyric acid (GHB), all of which can also block memories.

Heroin and morphine are classified as painkillers because they bind with the body's own opiate receptors, whose normal function it is to lessen pain in times of severe stress. From an evolutionary perspective this is a life-saving feature, as it allows animals to numb pain enough to be able to run and escape danger. It seems that one aspect of this natural pain relief is to stimulate the reward pathways so that the pain is masked with pleasure. Both heroin and morphine therefore act directly on this system by mimicking our own neurotransmitters; however, because heroin crosses the blood-brain barrier much faster, it causes a very intense rush of pleasure.

Hallucinogens, such as Lysergic acid diethylamide (LSD) and psilocybin (found in "magic mushrooms"), induce distorted experiences by mimicking serotonin in the pathways which are involved in perception. It is also thought that manipulating serotonin levels may directly influence consciousness, by blurring the lines between an awake-state and a dream-state. In the 1980s, a new and unique drug hit the nightclubs: Ecstasy, or MDMA (now often referred to as "Molly"), which leads to increased energy, euphoria and strong feelings of empathy. This drug is sometimes classified with the hallucinogenics and has a direct impact on serotonin levels; however, it is also a derivative of amphetamine, so it acts as a stimulant as well.

Finally, another substance that acts as a neurotransmitter imposter is cannabis, also referred to as marijuana, pot and weed. Much like the opiates, this chemical that occurs naturally in plants is very close in structure to a chemical that already exists in our body. Our knowledge of the body's own cannabinoid system is relatively recent, with cannabinoid receptors having been identified only in the 1980s. We still have a lot to learn about the functions of our natural cannabinoid system, but there is no doubt that the physiological and psychological effects of cannabis itself are quite diverse. On a positive note, it shows a lot of promise as a therapeutic agent in a number of medical conditions, such as cancer, asthma and multiple sclerosis.

Below: The opioid painkiller morphine occurs naturally in poppies. It mimics the body's own painkillers.

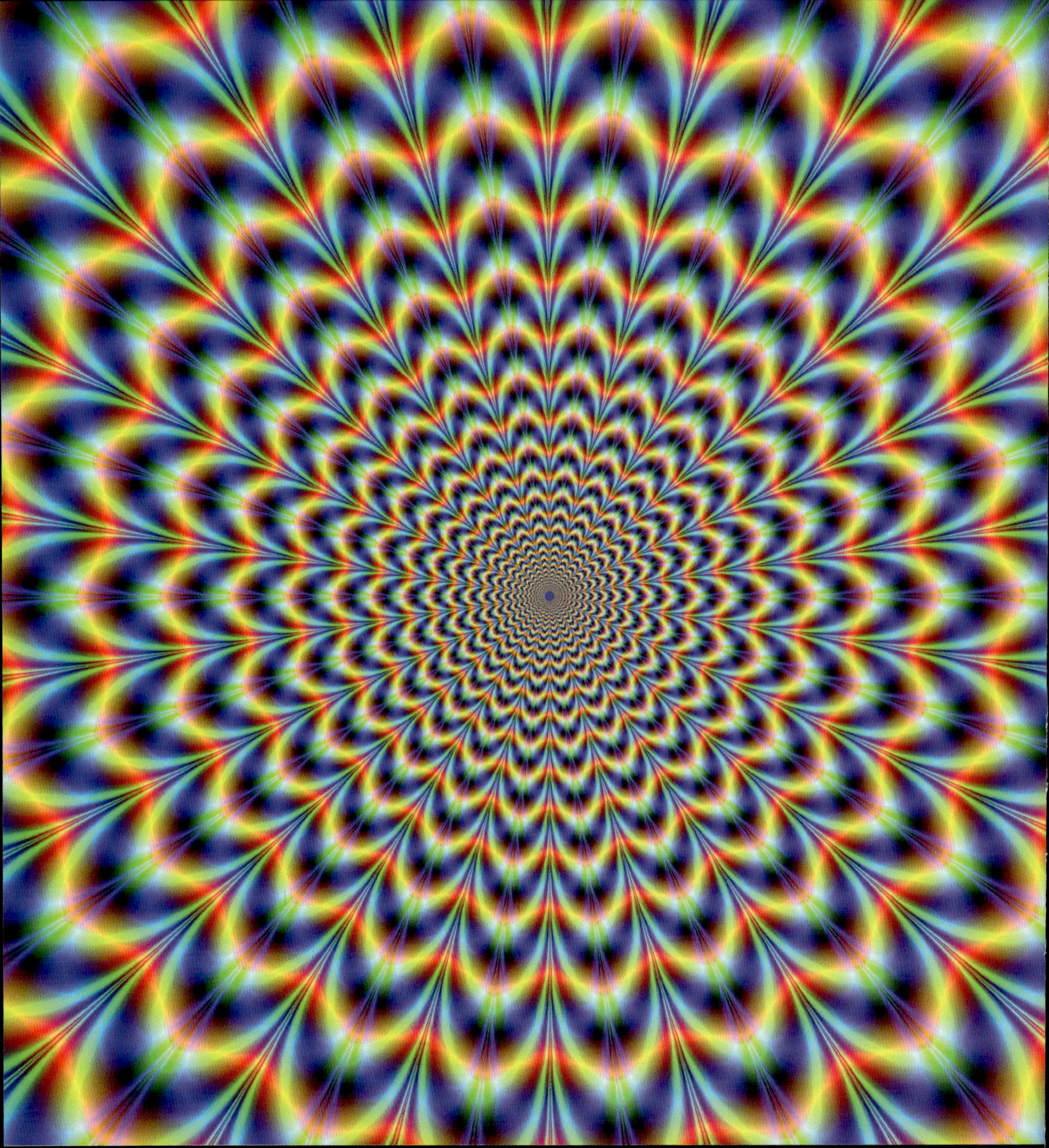

It seems, then, that there are a huge number of natural and man-made substances that can alter the brain's neurotransmitter systems. Some of them mimic, others block; some increase synthesis and others prevent breakdown. Whenever we interfere with the natural chemical processes in the brain, we risk upsetting a very complex and carefully balanced system. Nevertheless, drugs have been used for both recreation and medication for centuries and they have the potential to improve quality of life in many people. They also provide an important window into the psychological and biological mechanisms of mental health.

Opposite: Psychedelic drugs such as LSD and "magic mushrooms" alter cognition and perception.

Left: Cannabis is used as a recreational drug and as a therapeutic agent for conditions including multiple sclerosis.

MYTH BUSTER: THE CHEMICAL IMBALANCE THEORY OF MENTAL ILLNESS

A recent Australian survey found that 86 per cent of adults believe that mental illness is the result of a chemical imbalance in the brain. Many other studies confirm that this is a widespread belief across the world and the pharmaceutical companies often exploit this view. A television advert from the US chemical giant Pfizer states, "depression is a serious medical condition that may be due to a chemical imbalance", and claims that their drug Zoloft "works to correct this imbalance".

When the first antipsychotics and antidepressants hit the market in the late 1950s, they caused something of a revolution in the field of psychiatry because they seemed to have such dramatic effects on symptoms. Since these drugs were altering the levels of neurotransmitters, the natural conclusion was that they were normalizing the brain chemistry.

Although this theory was published and widely accepted at the time, there are obvious flaws in the idea. For example, aspirin is a very effective painkiller, but we do not say that headaches are the result of taking too little aspirin. Similarly, alcohol may help people with social anxiety, but most people would agree that this is masking the problem, rather than fixing it. There are other reasons to question this theory – it does not account for the timescale of antidepressant effects, nor does it explain why two drugs that have contrasting effects on serotonin both provide relief from symptoms.

Therefore, while drugs used in psychiatry may be extremely useful in alleviating distressing symptoms, it is certainly not as simple as saying that they are correcting a chemical imbalance. Nevertheless, we might still learn something about the underlying mechanisms: one promising new avenue of research suggests that antidepressants may help stimulate the growth of new neurons, which may account for their effects, as well as explaining the timescale.

4. THE STRESSED BRAIN

WHETHER IT IS WAITING IN THE WINGS TO GIVE A BEST MAN'S SPEECH, OPENING UP A HUGE, UNEXPECTED BILL, OR PREPARING TO WRESTLE WITH A CROCODILE, MOST OF US HAVE EXPERIENCED ACUTE MOMENTS OF STRESS AT SOME TIME OR OTHER.

You know the feeling – your heart starts to pound, your mouth dries up, you breathe more quickly, your muscles tense, you have a feeling of butterflies in the stomach and you may start to tremble. These physical reactions can be quite useful if you are about to run away from a crocodile and may even be of some assistance if you are about to perform one of Hamlet's soliloquies on stage. However, for many "stressful" events, these bodily reactions are not only unhelpful, they can be quite destructive – especially if there is no immediate physical release or psychological resolution of the situation. So, what is stress and why is it such a problem in our modern world?

Above: Most of us have experienced moments of acute stress at some time or another.

THE BRAIN'S TWO RESPONSES TO STRESS

Strictly speaking, the term *stress* describes any demand that is placed on the body, whether it is mental or physical. We tend to think of stress as being a negative experience, as in the situations described above. However, according to Hans Selye – the Hungarian-born "grandfather" of the field of stress research – there is a clear distinction between good stress (*eustress*) and bad stress (*distress*). Selye suggests that any challenge or change is a form of stress and that often this can be something that is fun and enjoyable, something that makes us feel alert, excited and driven. Of course, there are big individual differences in how we respond to different events and whether they lead to eustress or distress. For one person, having to deliver a best man's speech might be a dream come true – a chance to stand in the spotlight and show off their comedic talents – whereas another might feel physically sick at the thought of speaking to a crowd. The latter response is far more common and is, in fact, the most reliable way for scientists to induce a stress response: the *Trier stress test* is a widely used stress-induction procedure that requires participants to speak in front of a panel of people, while having their physiological responses monitored.

As individuals we also differ hugely in the *amount* of challenge that we are comfortable with. Are you someone who seeks out the highest and fastest ride at the theme park, or do you prefer a sedate tea party with a couple of close friends? Do you find that a bit of nervousness can help you

THE STRESSED BRAIN 53

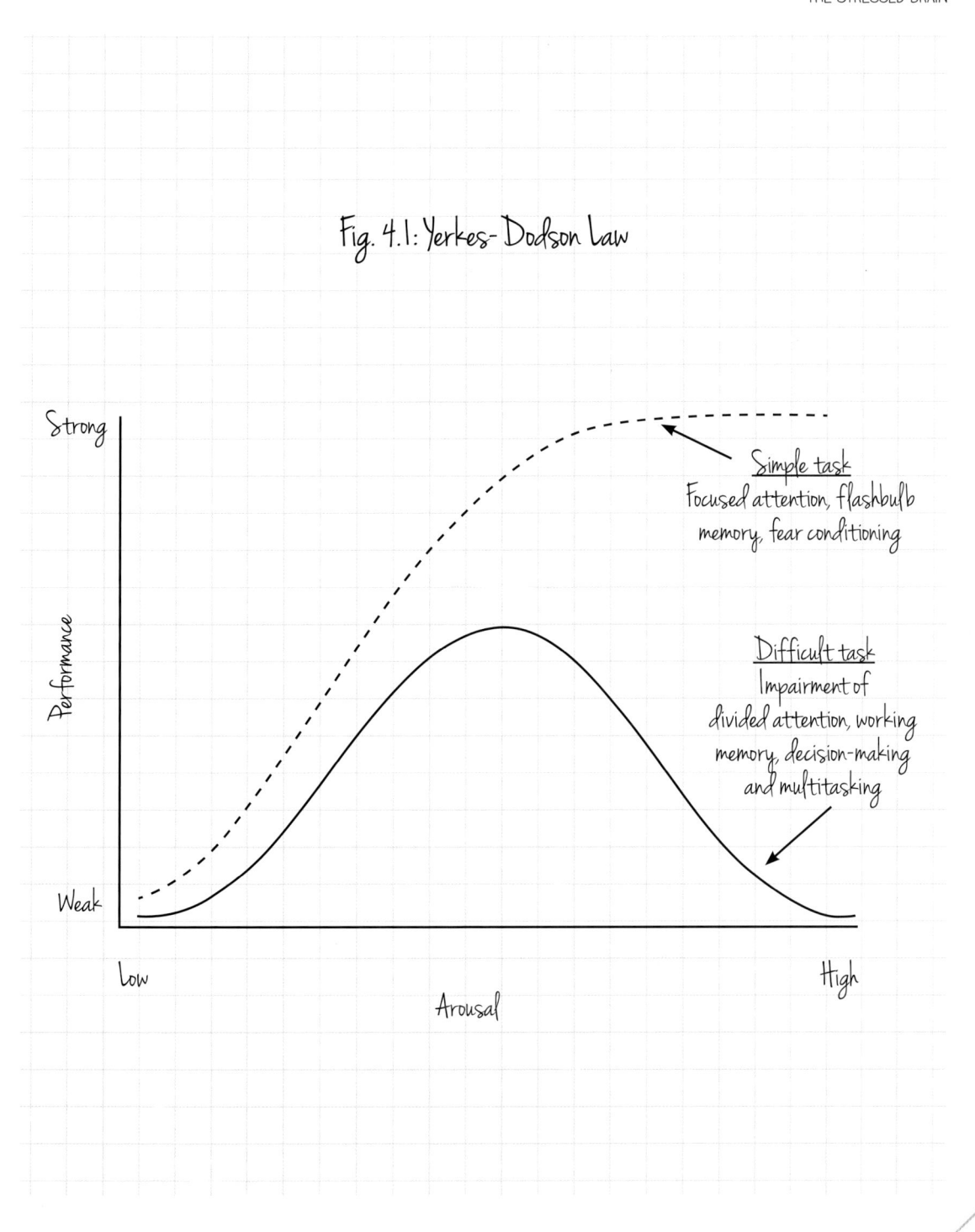

Fig. 4.1: Yerkes-Dodson Law

Simple task
Focused attention, flashbulb memory, fear conditioning

Difficult task
Impairment of divided attention, working memory, decision-making and multitasking

54 THE STRESSED BRAIN

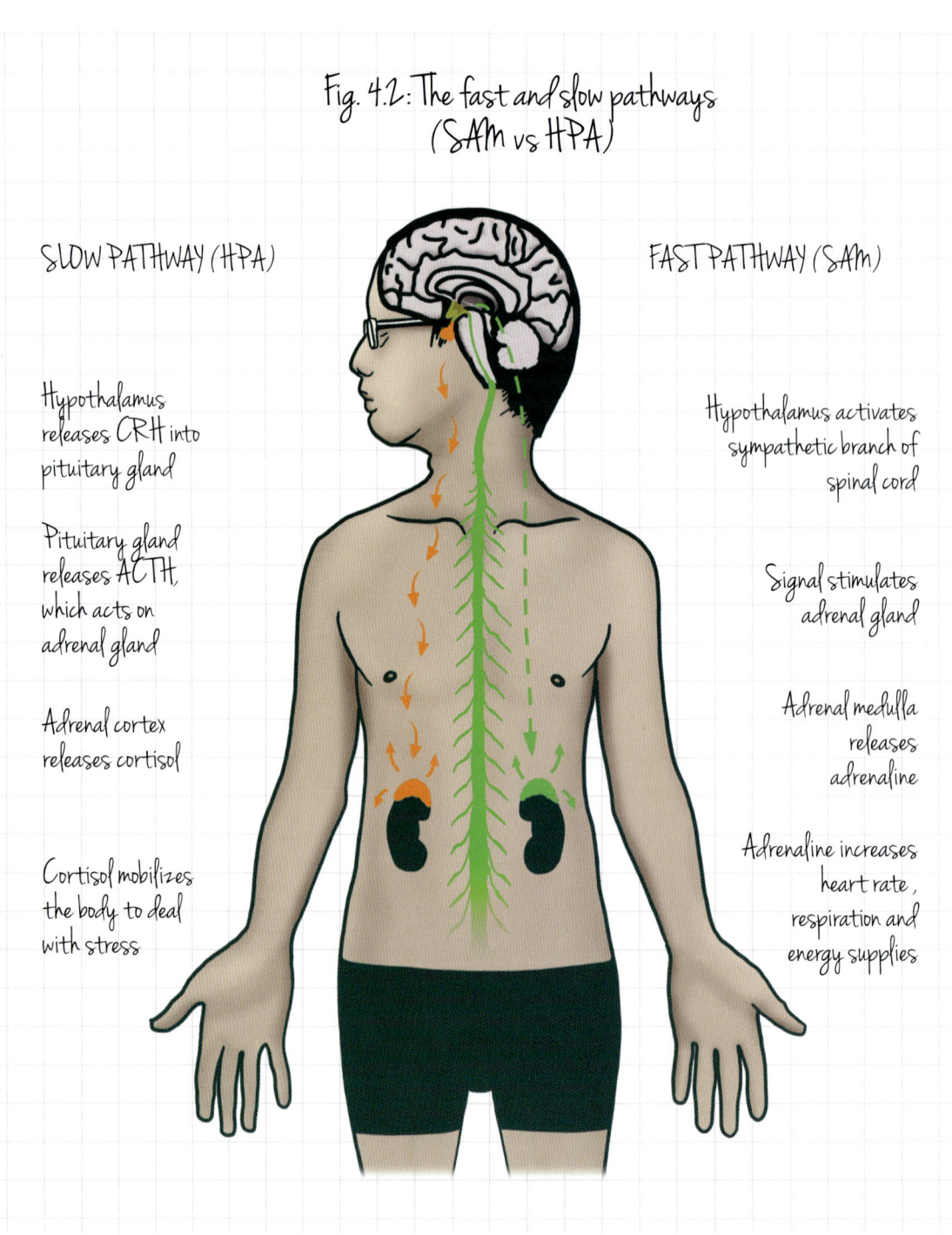

Fig. 4.2: The fast and slow pathways (SAM vs HPA)

SLOW PATHWAY (HPA)

Hypothalamus releases CRH into pituitary gland

Pituitary gland releases ACTH, which acts on adrenal gland

Adrenal cortex releases cortisol

Cortisol mobilizes the body to deal with stress

FAST PATHWAY (SAM)

Hypothalamus activates sympathetic branch of spinal cord

Signal stimulates adrenal gland

Adrenal medulla releases adrenaline

Adrenaline increases heart rate, respiration and energy supplies in an exam or tennis match, or do you go to pieces? A series of experiments at the turn of the century led American psychologists Robert Yerkes and John Dodsen to conclude that each of us has our own optimum level of arousal at which we perform best; anything below that will mean we are not driven enough, while anything too far above that will make us anxious and unable to function properly. Yerkes and Dodsen described this effect using the Yerkes-Dodsen law (*see* Fig. 4.1.), and it explains why some people thrive on leaving everything until the last minute and others cannot bear the anxiety of a looming deadline.

Stress, whether it is good or bad, is one of the most palpable demonstrations of the connection between mind and body. Just thinking about an upcoming interview or performance is enough to cause that familiar "butterflies in the stomach" feeling. In Chapter 7 we will examine in more detail the physical manifestations of different emotional states, but for now let us focus on the very specific physiological response we have to a *stressor* (any event that provokes stress). Stressors come in all shapes and sizes: some events are very sudden and require an immediate reaction (for example, seeing a child about to run out in front of a moving car); others present a more ongoing challenge and call for a more prolonged response (such as a bereavement or moving house). Luckily, humans have developed two different systems – one described as "fast" and the other as "slow" – which together give us remarkable flexibility when it comes to dealing with different stressors.

The "fast" pathway is often referred to as our *fight-or-flight* response, or more formally, the Sympathetic Adrenomedullary (SAM) system. When we perceive an emergency, such as a child running into the road, a signal is passed to the hypothalamus, the small region in the middle of the brain that controls the Autonomic Nervous System (ANS, *see* Chapter 1, page 12). The sympathetic branch of the ANS sends nerve impulses down to the adrenal glands, just above the kidneys, which in turn release the hormone adrenaline. Once in the blood stream, adrenaline sends immediate, targeted messages to different parts of our body, preparing it for urgent action – the heart and respiration rate speed up, providing more oxygen, the liver releases energy to the muscles, and our digestive system shuts right down,

which is why we may have a dry mouth or not want to eat. In moments of extreme and sudden stress, people sometimes describe an electric-shock feeling across their body, which gives some indication of how quickly this process kicks in.

The sudden boost of adrenaline produced by the SAM system is ideal for getting us out of sticky situations in a hurry, but it is a short-acting solution: under normal conditions, the levels fall again as soon as we believe we are out of danger and the body will return to a more relaxed, resting state. Whenever there is a stressor that lasts for more than a few minutes, the slower and longer-acting Hypothalamic-pituitary-adrenal (HPA) system kicks into action. Once again the stressor is processed in the brain by the hypothalamus, but in this case it causes the direct release of a hormone into the bloodstream, via the pituitary gland

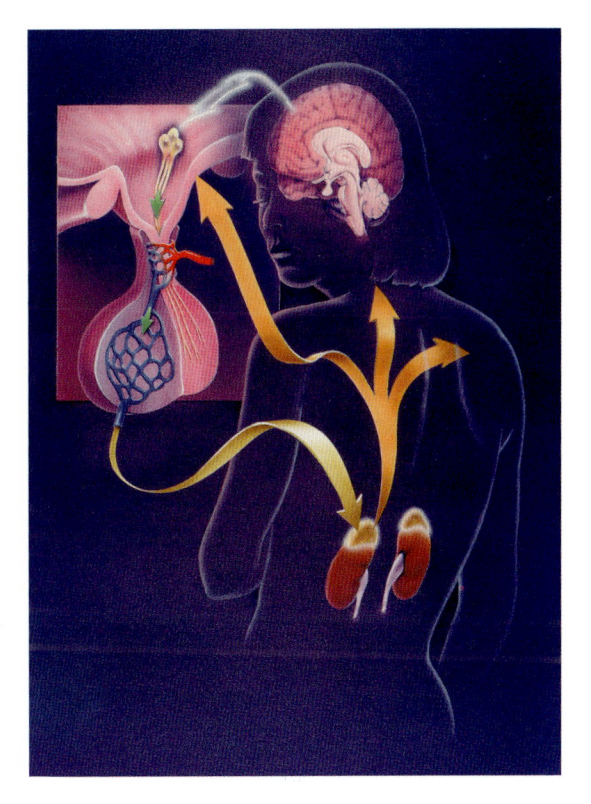

Above: The pituitary is the "mouthpiece of the hypothalamus", and controls the release of all hormones.

(often described as the "mouthpiece" of the hypothalamus). This sets off a cascade of chemical events (see Fig. 4.2) but the important end point is a release of the hormone cortisol, also via the adrenal gland. Cortisol plays a very important role in the body, regulating many functions from glucose levels and blood pressure to the immune response, bone density and even memory. And it is this fascinating little hormone that makes us so good at staving off colds during a busy period, and yet so vulnerable to catching them when the holidays come.

COMMUNICATING WITH HORMONES

Before we look in more detail at the all-important cortisol, let us take a short diversion to consider more generally what hormones are and how they work. The fast and slow stress pathways provide an excellent illustration of the two distinct ways in which the brain controls the body – the nervous system and the endocrine system. As we have seen in earlier chapters, neurons transfer information from one part of the brain or body to another, using electrochemical nerve impulses. This is a fast and direct form of communication, which is perfect for, say, picking up a cup of coffee or kicking a football. In contrast, the endocrine system exerts its influence on the body by releasing hormones into the bloodstream. This means that it takes a little longer for messages to reach their destination and they generally have a more widespread reach, locking onto receptors in a number of different organs. To give one example, the female sex hormone oestrogen influences the uterus, the heart, breast development, fat production, blood vessels, bone formation, salt regulation and brain function, as well as playing a key role in pregnancy and birth.

In essence, the way in which the nervous system operates can be likened to making a direct phone call or sending an email. Conversely, the endocrine system works in a way that is more akin to transmitting a message over the radio or communicating via social media – the messages are out there to be picked up by anyone to whom they have relevance. While the nervous system can coordinate incredibly complex and refined motor sequences and enable us to process highly detailed visual scenes or musical performances, the endocrine system manages longer-term

Above: The nervous system makes direct one-to-one contact with specific parts of the body, much like a telephone.

Right and opposite: The endocrine system sends hormones around the body in the bloodstream, where they are picked up by the relevant organs. It is more like a speaker system.

THE STRESSED BRAIN 57

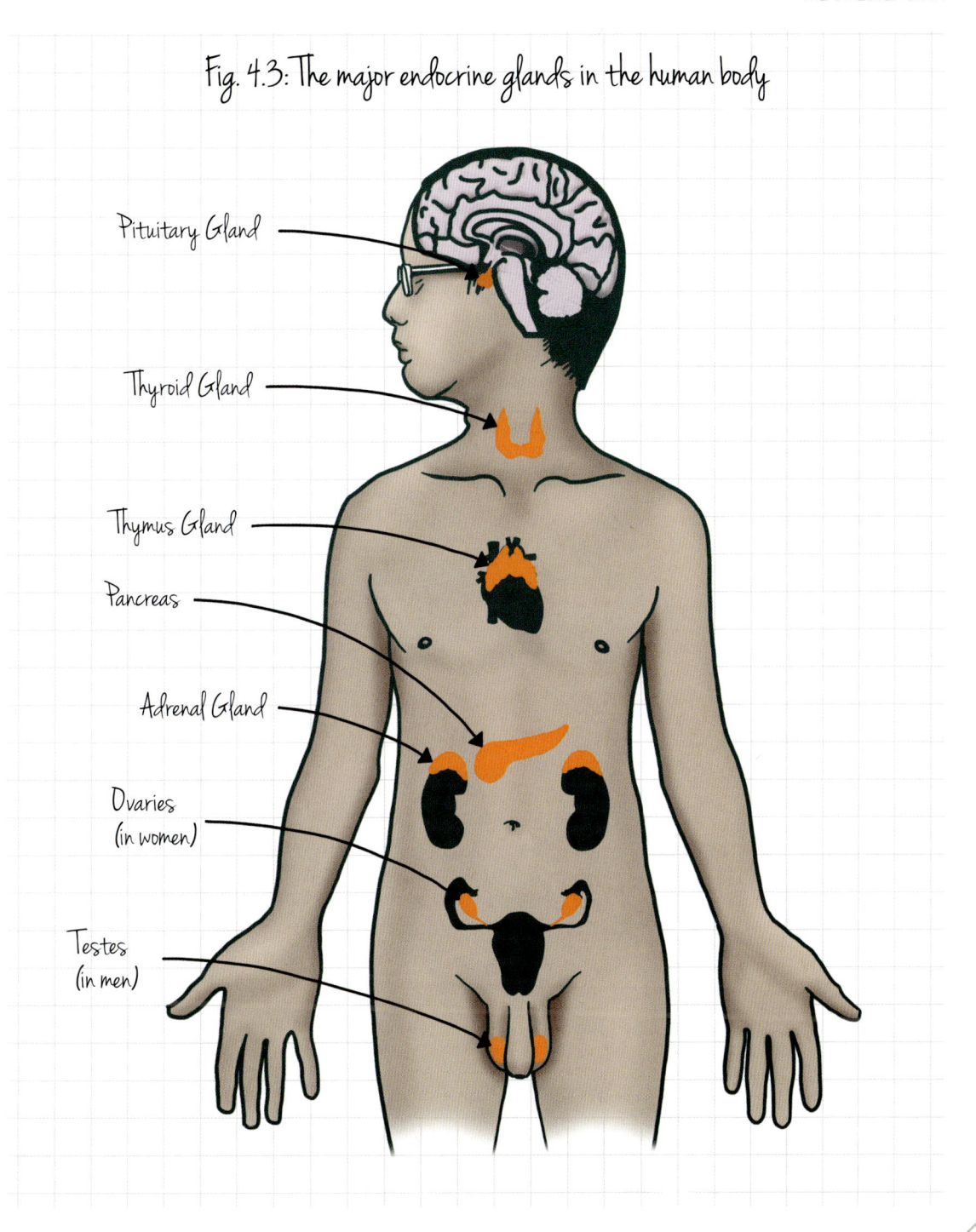

Fig. 4.3: The major endocrine glands in the human body

- Pituitary Gland
- Thyroid Gland
- Thymus Gland
- Pancreas
- Adrenal Gland
- Ovaries (in women)
- Testes (in men)

processes: reproduction; growth; response to stress; and the maintenance of a stable internal environment in an ever-changing world.

All hormonal actions are orchestrated and led by the hypothalamus. This means that damage to this area can have fatal consequences – levels of sugar, oxygen, salt and even temperature are all kept within safe limits through hormonal control. The pituitary gland, which hangs down from the hypothalamus, is described as the master gland, because it releases a wide range of hormones, many of which trigger the release of other hormones from endocrine glands distributed throughout the body (*see* Fig. 4.3). These little messengers then swim around the body in the bloodstream, where they can directly or indirectly influence other parts of the body, including the brain itself. Insulin, for example, is a hormone released from the pancreas to keep levels of glucose within safe limits; however, when this system fails, as it can in someone with diabetes, it may have serious consequences for the brain. A sudden catastrophic drop in glucose levels can lead to coma or even death, whereas longer-term disruptions have been linked with accelerated ageing of the brain and dementia.

CORTISOL

All this brings us neatly back to cortisol, the "stress hormone" that we discussed briefly earlier. This glucocorticoid hormone has a particularly significant impact on the brain – dysfunction has been linked with memory impairments, as well as depression, "burnout" and psychosis. So what does cortisol actually do?

In actual fact, to describe cortisol as a "stress hormone" does it a disservice. In essence, it enables our body to respond and adapt to change in the very general sense and it regulates many of our daily physiological functions – glucose, electrolyte levels (for example, sodium and potassium), liver, kidneys, blood pressure and wound healing. In healthy people, cortisol follows a very predictable daily routine and is a vital part of maintaining a natural circadian rhythm (*see* page 102). Levels of cortisol increase when we wake up in the morning, driving our brain and body to be ready for the demands of the day ahead; levels then gradually reduce throughout the day, gradually rising again after a few hours sleep.

That first waking increase in cortisol – the Cortisol Awakening Response – has been found to be a very

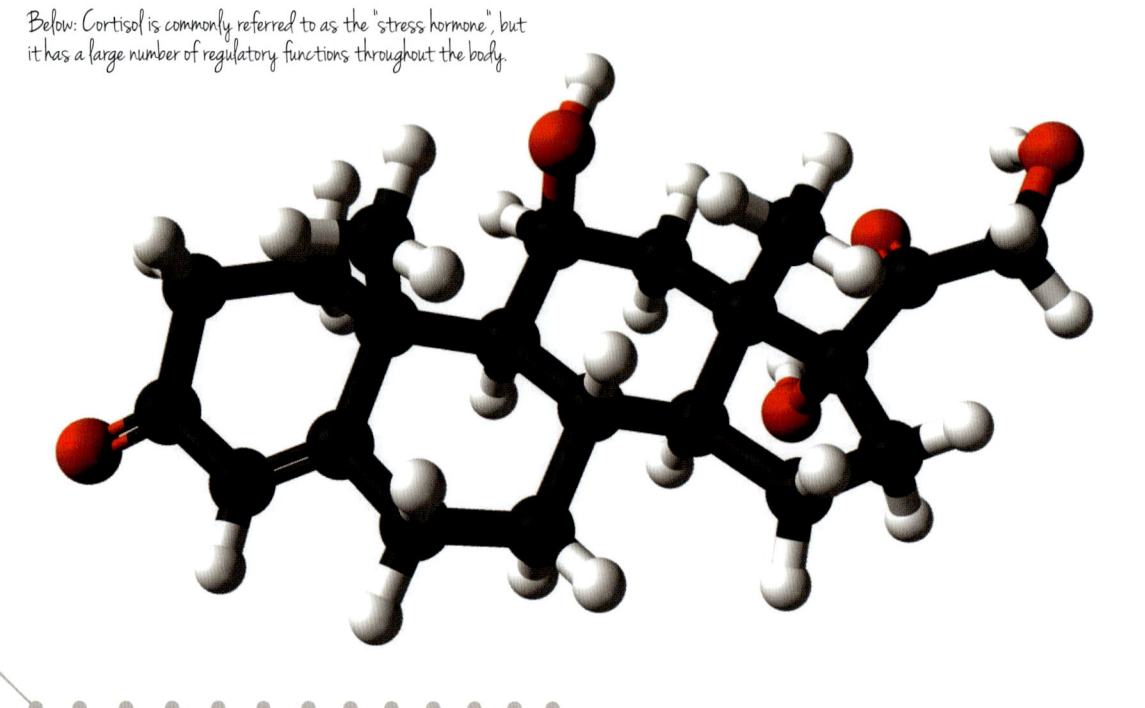

Below: Cortisol is commonly referred to as the "stress hormone", but it has a large number of regulatory functions throughout the body.

Above: Cortisol peaks first thing in the morning in a healthy individual and this is enhanced by natural daylight

significant indicator of health and wellbeing, and is often abnormal in people with type 2 diabetes, chronic fatigue syndrome, multiple sclerosis, "burnout", amnesia, depression, eating disorders and post-traumatic stress disorder (PTSD), to name but a few. There are even large studies that suggest the Cortisol Awakening Response predicts overall health and survival from cancer and there is little doubt that it is a key player in the link between our mental state and physical health. Angela Clow, a psychophysiology expert from the University of Westminster, has spent many years studying this phenomenon and has found that it is negatively affected by many things: acute and chronic stress, sleeping patterns, a heavy work schedule, fatigue, shift-work, pain and noise at night. On the other hand, she has shown that we have a healthier waking response when we get up early, are exposed to daylight, have a regular routine and exercise regularly. People also tend to have a bigger morning peak in cortisol when there is an exciting or demanding day ahead, although some fascinating recent research by Mark Wetherell at the University of Northumbria has shown that levels do not fall again, as we might expect, if the anticipated event does not go ahead.

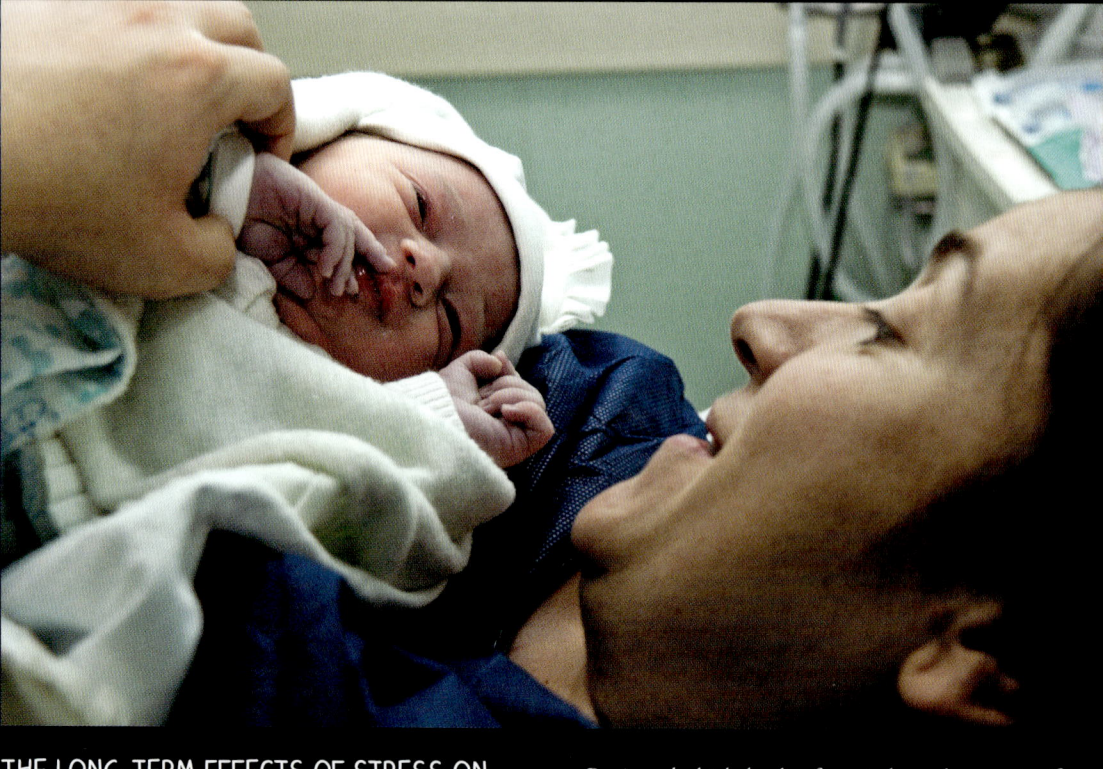

THE LONG-TERM EFFECTS OF STRESS ON HEALTH AND THE BRAIN

A healthy rise and fall in the hormone cortisol is crucial in enabling us to respond appropriately to our environment. Thus, while cortisol follows a basic regular daily rhythm, it also fluctuates throughout the day, depending on what we are doing. Extra bursts of cortisol are released whenever we eat or exercise – again a response to change – or levels will shoot up dramatically if we experience a stressful event, ready to provide back-up for the fast adrenaline response. Importantly, levels of both adrenaline and cortisol should drop again as soon as we have adjusted to the change or responded to the stressor. However, as many of us know, the root cause of stress cannot always be quickly or easily dealt with, and so our body maintains a state of readiness and high alert. That might not sound like such a bad thing on the face of it – always being on our toes and prepared to deal with anything the world might throw at us – but this comes very much at the expense of our long-term health.

Persistently high levels of cortisol can have a significant impact on many aspects of health and wellbeing, because this simple chemical messenger regulates such a range of important functions. In particular, cortisol plays a complex role in controlling the immune system, which means that prolonged elevated levels can compromise our ability to fight infections. Excessive cortisol has also been linked with diabetes, weight gain, cardiovascular disease, fertility problems and dementia. The developing brain seems particularly vulnerable: Vivette Glover, a British Professor of Perinatal Psychobiology, has shown that a mother who is particularly anxious or stressed during pregnancy is more likely to have a child with emotional problems, Attention Deficit Disorder (ADD), or delayed cognitive development. While Glover acknowledges that there are many influencing factors, her research reveals a specific measureable effect of cortisol. Interestingly, she has found that a stressful maternal experience during pregnancy might be more significant to a child than the effects of its own experiences soon after it is born.

can be, there is even evidence that severe and chronic stress during early childhood (for example, abuse or neglect) may be partially responsible for memory problems in middle to old age. In fact, there is a powerful and complex relationship between cortisol, memory and learning. In summary it seems that small bursts of cortisol can be quite helpful in cementing memories, but once again prolonged high levels have a negative impact. This seems in part to be because cortisol has quite toxic effects on the hippocampus – an area of the brain that enables us to lay down memories (*see* Chapter 5). There is something of an irony here, because this same region of the brain is normally responsible for keeping track of cortisol levels and sending a signal to the hypothalamus to hold back when levels are becoming too high. Therefore, continued high levels of this important hormone can eventually destroy the very brain cells that are there to keep it in check, thus allowing it to free-wheel.

Opposite: Cortisol plays a vital role in the development of a healthy foetus, but very high levels in the mother can have a negative impact.

Below: Small bursts of cortisol can help to cement memories, but when levels are high for a prolonged amount of time it can impair learning.

Ultimately this can lead to very poor regulation of cortisol, which, as we have seen, has serious consequences for mental and physical health.

The negative impact of stress does not stop with cortisol: continued activation of the faster emergency SAM system also takes a significant toll. The Autonomic Nervous System has two branches that have opposing actions – the Sympathetic Nervous System, which we have seen activates a flight-or-flight response by releasing adrenaline, and the Parasympathetic Nervous System, which supports the long term good of the body and enables us to "rest and digest". These two systems are like a see-saw and, because of the way they work, they cannot be active at the same time. So while the sympathetic branch increases heart rate and slows digestion, its parasympathetic opponent slows the heart down and activates the digestive process, in an effort to absorb and preserve energy. This is why your stomach may start rumbling when you are feeling particularly relaxed. By definition, whenever the sympathetic nervous system is activated, it is at the cost of all the day-to-day physiological processes that keep us healthy and well.

Hopefully, then, it is quite obvious that long-term chronic negative stress – distress – is not good for us. But we must no

Above: There are many practical ways to actively lower levels of stress hormones.

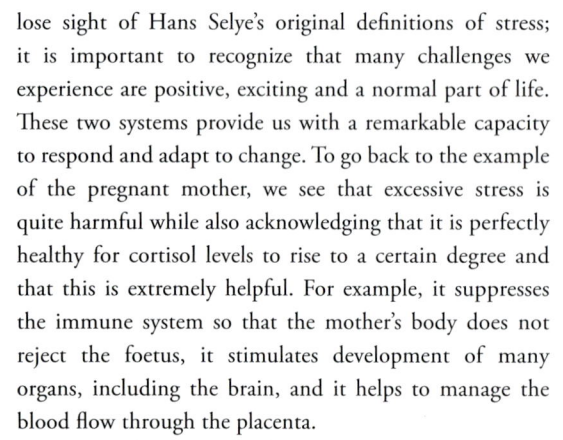

lose sight of Hans Selye's original definitions of stress; it is important to recognize that many challenges we experience are positive, exciting and a normal part of life. These two systems provide us with a remarkable capacity to respond and adapt to change. To go back to the example of the pregnant mother, we see that excessive stress is quite harmful while also acknowledging that it is perfectly healthy for cortisol levels to rise to a certain degree and that this is extremely helpful. For example, it suppresses the immune system so that the mother's body does not reject the foetus, it stimulates development of many organs, including the brain, and it helps to manage the blood flow through the placenta.

Stress, in the most general sense, is an important element of day-to-day function – it motivates and excites us, as well as preparing us physically for anything the world can throw at us. The all-important thing is for us not to stay in that "red-alert" phase for too long, but to find ways in which we can activate our parasympathetic system and allow our cortisol levels to return to their natural level. We cannot always remove a stressor, but there are many active steps we can take to reduce the physical and mental effects: for some, it may be as simple as taking a gentle walk through the park; for others, it may be practising yoga or meditation, having a long quiet bath, connecting with good friends, finding something that makes them laugh, or listening to a favourite piece of music. All of these things have been shown to significantly reduce the physical manifestations of stress. An alternative approach is to seek support in changing perceptions – we treat something as a stressor only if we perceive it to be one, so reframing a situation can be very helpful. A final simple and practical solution is to respond to the call for fight-or-flight by doing some intense exercise: go and expend all the additional energy and resources that your body has generously given you to fight that crocodile!

TOP TIPS: GETTING GOOD SLEEP

Sleep is still not well understood, but we do know that it is critical to good brain function, so it is important to get good quality sleep. Sleep plays a vital role in many things – for example, cementing our memories, helping us to solve problems and make decisions, and clearing out toxins. A poor quality of sleep has been statistically linked with poorer health, faster ageing, higher rates of depression, more forgetfulness, obesity, lower alertness and more accidents. So, how can we ensure a good night's sleep?

1. Have a regular routine – our bodies follow a natural circadian rhythm and respond well to habits;
2. Make sure that lights are dimmed for an hour before sleep (including no computer or television screens), and that you have complete darkness during sleep and as much light as possible during waking hours. This is because light and dark determine levels of cortisol and melatonin, which drive our sleeping and waking;
3. Take regular exercise – this helps our brain and body to relax (but avoid anything too energetic just before sleep);
4. Reduce your intake of alcohol, caffeine and other psychoactive drugs – these disrupt the sleep-wake cycle;
5. If you are a natural worrier, try techniques that help you relax, such as yoga or meditation. We are programmed to stay alert and awake when anxious. Keep a pen and paper by the bed to write down things that are worrying you, so that you can let them go.

Most importantly, if you really cannot sleep, then get up for a little while and try again a bit later. If all else fails, bear in mind that the occasional bad night is nothing to be concerned about.

5. HOW THE BRAIN MAKES MEMORIES

TRY TO REMEMBER AN IMPORTANT EVENT IN YOUR EARLY LIFE, SOMETIME BETWEEN THE AGES OF 5 AND 10 YEARS. STOP READING FOR A MOMENT AND IMAGINE IT IN AS MUCH DETAIL AS YOU CAN.

Now, think about these questions: how vivid is the memory? Can you remember what you were wearing, what the weather was like, whom you were with, what time of day it was, how you were feeling? Are you looking out at the scene as if through your own eyes or are you looking down upon it as an observer? If your mother, father or other relative was there, do they look like they looked then or do they look like they look in more recent times? What do you notice about what it feels like to remember? What happens if you repeat all of this for a much more recent event? These are just some of the questions that psychologists ask when they are trying to understand what memories are and how they are formed.

WHAT DOES IT MEAN TO REMEMBER?

We tend to think of our past in the same kind of way we think of a film, as a flowing narrative of events that naturally follow on from each other. Memories provide a "back story", a context that enables us to make sense of the present. However, when we examine memory in detail we find it is far more disjointed – a patchwork of sights, sounds, smells, thoughts and feelings that we piece together into cohesive moments. These moments are collected together to make up memories of discrete events and then presented alongside each other in our perceptual world to give us a sense of coherence and continuity.

Each time we remember a moment from our life we construct it anew. We do this by using the building blocks of *episodic memory* – recollective feelings of being somewhere; and *semantic memory* – concrete knowledge about our world (for example, that snow tends to occur in the winter months) and about our personal history (such as where we went to school). The first of these, the experiential part of remembering, can only ever be a record of what we have perceived and understood to have happened, and so will always be dependent on the individual viewpoint (physical, emotional or cognitive). The context in which memories are recalled is also important and differs from one person to another. Because of this, two people's memory of the same event might differ quite dramatically, but each is equally valid. Memories are almost never 100 per cent accurate, and all memories are malleable and changeable.

What is the evolutionary benefit of this? Why would we have a memory system that is not entirely accurate? The reason, experts believe, is because a key purpose of memory is to provide us with a sense of coherence, a way of connecting events in the world to our internal representations and giving us a sense of continuity. It is usually important to have accuracy in terms of semantic knowledge, such as where we live, but when it comes to episodic memories – memory of our experiences – accuracy is less significant than consistency. In order to have a strong sense of self, it is important that our memories fit with our beliefs and feelings about ourselves.

The idea that memory is central to our sense of who we are is supported by the fact that we have particularly strong recollections of events that happened during our adolescence and young adulthood, a period of time that is crucial in identity formation. When people over the age of 40 are asked to name their favourite books, films, music, or footballers, they gravitate towards events that they experienced between the ages of 15 and 30. This is an incredibly consistent finding and has become known as the *reminiscence bump*. One theory for this phenomenon is that people mentally return to this period of time more often than other stages of their lives because it is full of *self-defining moments* – times when a lot of life-shaping decisions are made.

LOSING MEMORY

The twentieth-century Spanish film director Luis Buñuel once said, *you have to begin to lose your memory, even if only in bits and pieces, to realize that memory is what makes our lives.* This is probably true and is sadly something that too many individuals have to face. Around 36 per cent of people who survive a serious head injury will be left with amnesia, more than 10 per cent of people over the age of 65 have dementia and more than 70 per cent of people with AIDS will suffer from memory impairments. Other causes of memory loss include encephalitis, strokes, multiple sclerosis, Parkinson's disease, hydrocephalus, epilepsy, meningitis, brain tumours and depression.

Memory impairments come in many forms and always impact heavily on quality of life. Two particularly devastating presentations are the dense amnesia that can occur after brain damage or infection and the progressive memory loss that develops in Alzheimer's disease. The two most famous cases of amnesia are an American patient named Henry Molaison, more widely known as HM, and Clive Wearing, a highly successful British musicologist and conductor. For many years HM had suffered from a particularly bad case of epilepsy, which caused huge disruption to his life. This became so debilitating that his neurosurgeon decided it was worth taking the drastic step of removing his temporal

Below: Self-defining experiences in our youth often form our most powerful memories.

lobes, where the seizures were starting. The operation was a success in terms of controlling the epilepsy, but HM was left with amnesia and, although he was able to remember much of his past, he was unable to form any new memories. He was extensively studied in the 1950s by the Canadian neuropsychologist Dr Brenda Milner and continued to work with psychologists until he died at the age of 82, 55 years after his operation. Clive Wearing (described in more detail in Chapter 9, page 119) developed an even more extreme form of memory loss, following an encephalitis infection. Clive, who is now in his seventies, is probably the most extreme case of amnesia ever recorded. He has no recollective memory whatsoever of anything that has happened more than fifteen to twenty seconds previously.

THE VITAL LEGACY OF HM AND CLIVE WEARING

There are a number of important things to say about both Clive and HM. The first is that despite their amnesia they both retained full awareness of their personal identity – their names, basic biographical details such as who their family was, and where they went to school. Contrary to the stories that are often presented in films (and which a recent study found were believed by more than 80 per cent of the general public) people with an organic amnesia do not forget who they are. There are a few situations in which this can happen, but generally this is when someone is experiencing what is described as a *fugue state* or a *dissociative amnesia*, in which the memory loss is brought on by psychological trauma as opposed to brain injury.

There are other important things that we have learned from HM and Clive Wearing, which have also been consistently found in other people suffering from amnesia. The first is that some aspects of memory seem not to be affected at all – for example, both Clive and HM maintained an ability to repeat back fairly lengthy lists of numbers. They also both showed signs of *implicit* memory, that is, their behaviour was shaped by their experiences, despite them not having any conscious memory of what they had seen or heard. More recent studies have shown that it is possible to teach people with amnesia quite complex tasks, such as riding a bike or touch-typing, and they will hang on to those skills, despite not remembering ever having learned them. Importantly, both Clive and HM retained their intellect and their ability to engage in conversation, albeit with some repetition, which highlights the fact that memory is not a prerequisite for intelligence.

One of the most striking things about working with people with amnesia is the huge significance of memory in day-to-day living. It enables people to create and maintain a sense of identity and to engage in social relationships, and it also provides the basis for imagining and planning the future. Every conversation we have and every decision we make relies on reference to our past experiences. Nowhere is the value of memory more noticeable than in people who develop an Alzheimer's-type dementia. The earliest symptom manifests itself as a difficulty with laying down new memories; for someone in the first stages of Alzheimer's, things that are said and done are akin to writing words in the sand – they can be quickly washed away. As the disease progresses, it becomes increasingly difficult for the individual to follow stories and navigate around familiar places. Eventually, the disease begins to wipe out longer-term memories, until even closest relatives are no longer recognizable. Any form of amnesia is devastating, but the complete erosion of personal history that is seen in people in the latter stages of dementia is very distressing and utterly incapacitating.

Below: For those in the early stages of Alzheimer's, memories can be quickly washed away.

From a neurological point of view, there are two key things we can learn by studying people who have memory loss. The first is that there are some areas of the brain that are critical for laying down and retrieving memories: Clive, HM and people suffering from Alzheimer's disease have all been shown to have pathological changes in the *hippocampus* – a seahorse-shaped structure located in the temporal lobes of the brain. In addition, people with dementia begin to lose their longer-term memories as the damage spreads to the rest of the brain and across the cerebral cortex. This phenomenon suggests that, while the hippocampus might be important in making memories, they are actually stored throughout the brain.

A second thing we can learn from Clive and HM is that memory exists in a number of different forms: while some aspects of remembering are affected by the damage to the brain, others seem to be more resilient. Different brain injuries cause patterns of deficits that are quite unlike those described in typical amnesia. For example, patients with damage to their cerebellum may no longer remember how to ride a bike but have no problems recollecting what they had for breakfast or the name of their favourite band. These neuropsychological findings provide strong evidence that memory is a complex and multi-faceted construct. This idea is backed up when we look at the brain activity of people undertaking different memory tasks. So, let us look at exactly what this means in practice.

NON-UNITARY THEORIES OF MEMORY

The notion that memory is an intricate skill with many separate dimensions seems to make intuitive sense. Consider the archetypal absent-minded professor, who is like a walking encyclopaedia and yet forgets where he has put his car keys and cannot remember to post a letter. We all have aspects of memory that we are good at and some that we are not so good at – maybe we are quick to learn a new skill or a list of numbers but are always forgetting our friends' birthdays. As well as the lessons we have learned from the likes of Clive and HM, we now have an abundance of evidence that supports this non-unitary notion of memory.

The most clear-cut distinction is that between conscious memory, sometimes called *declarative* memory, and unconscious (*non-declarative*) memory. Declarative memory refers to anything to which we have conscious access – our autobiographical memory, a list of phone numbers, the knowledge we acquire. However, much of our everyday behaviour depends on non-declarative memory. Loosely speaking, this relates to any situation in which our previous experience influences the way we behave, without us being aware of it. An obvious example of this is the ability to drive a car or ride a bike. In the early stages of learning these skills we have to refer constantly to what we have been taught, but those abilities very quickly become automatic and habitual. There are many other examples of unconscious memory, including having emotional responses to certain situations without knowing why, or being unknowingly influenced by advertising.

There is also convincing evidence that short-term and long-term memory are two distinct processes that rely on different parts of the brain. In everyday conversation we tend to use the term *short-term memory* to refer to all recent events – for example, what we had for breakfast this morning – and we use *long-term memory* to cover our distant past. However, for psychologists, *short-term memory* means anything that we hold in mind temporarily but lose as soon as we become interrupted – good examples might be counting money or doing mental arithmetic. Anything else is considered by psychologists to be part of long-term memory, even if the event occurred only ten minutes ago.

What about that absent-minded professor who forgets to post a letter or lock the back door? The ability to avoid such behaviour – that is, to have a reliable memory for things we must do – is described as *prospective memory*, and describes our ability to carry out particular tasks at the correct time. This includes things like remembering to take your medication or picking up a fresh pint of milk next time you pass the supermarket. Forgetfulness for these everyday tasks is a complaint shared by many of us. However, for some people not having a good prospective memory can actually become quite dangerous – for example, if they forget to turn off the gas stove or take their medication on time. People that have damage to the prefrontal cortex seem to have particular problems with this aspect of remembering, despite being quite good at others.

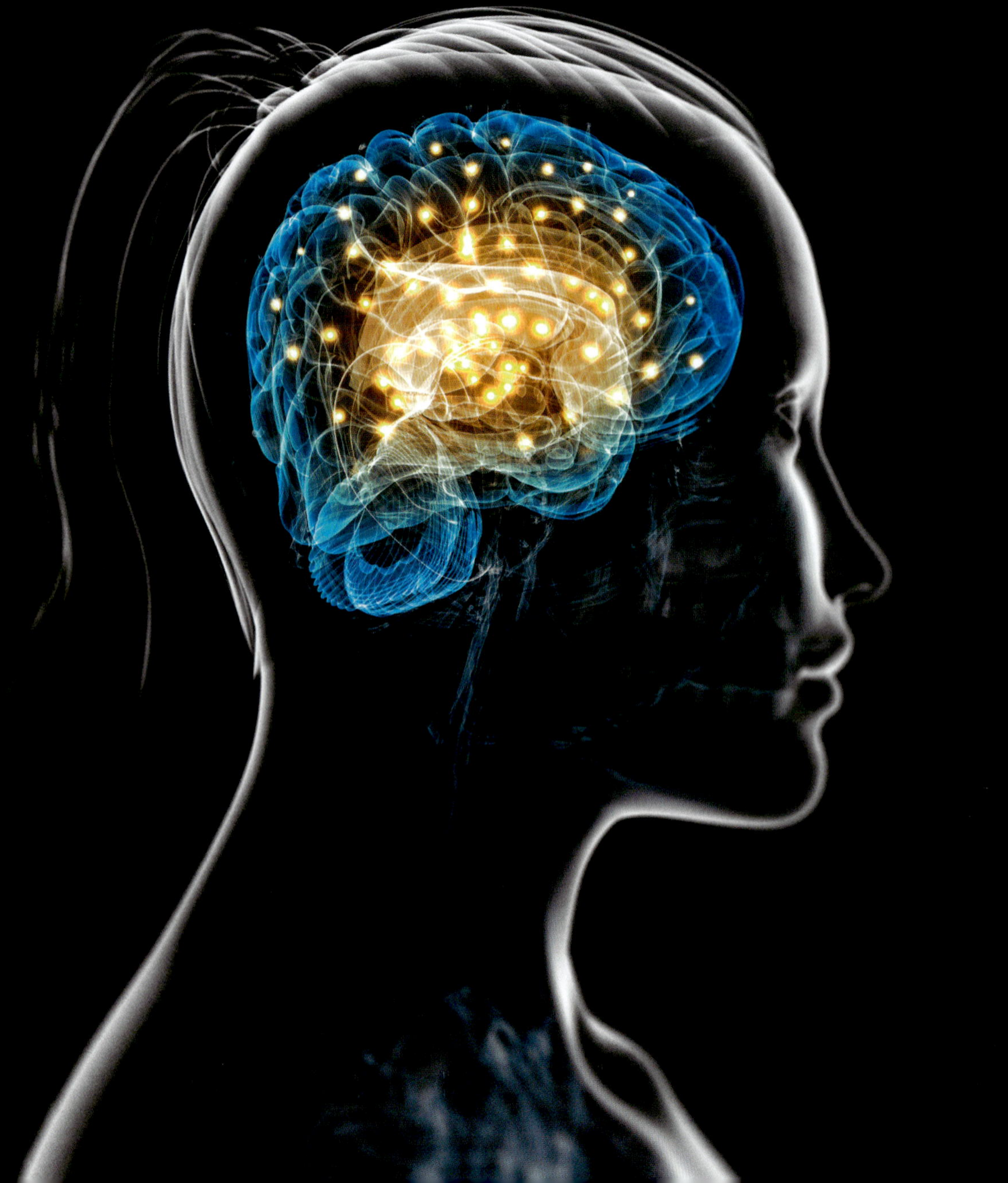

BUILDING CONNECTIONS

What does any of this tell us about how memories are formed in the brain? To answer this question we need to go right back to the turn of the twentieth century and a man named Karl Lashley, who was determined to find the location of what he called the *engram* – an individual memory trace. Lashley, an American psychologist and behaviourist, spent most of his life teaching rats how to navigate a maze and then systematically removing a different part of their brains with the aim of trying to remove their memory. Despite ultimately working his way through the whole of a rat's brain, using countless different rodents, he failed in this mission. The one thing he did discover was that the more of the rat's brain he took out, the worse the memory was.

Lashley had a student named Donald Hebb, who continued with this quest to locate memory. His work led him to publish an important conclusion in 1949: memory is not stored in a single place in the brain, but instead is stored across the cerebral cortex as networks of neurons. He called these *cell assemblies*, and demonstrated conclusively that repeated associations being made between different events create them. Thus, being regularly exposed to the smell of cut grass during the summer months means that the smell becomes associated with a warm sun, long days and summer holidays.

A key premise of Hebb's work was that the more frequently a synapse was activated, the more easily it would be triggered in the future. This can be likened to walking through a field of tall grass – the first time is hard work, the second time is a little easier and by the time it has been done twenty times, there is a clear pathway that can be crossed effortlessly. Hebb's theories have often been summarized as *neurons that fire together, wire together* and is still seen to be the key mechanism by which we form memories.

Modern evidence suggests that each time we go back over an event in our mind we are strengthening these connections and bedding in the memory. Sometimes we actively choose to remember, but at other times memories just pop into our mind because we come across a cue – a sight, smell or sound – which triggers a memory by reactivating the cell assemblies. Hebb's notion that learning occurs because of changes to the synapses was one of the

Opposite: Memory is not stored in a single place in the brain, but across the cerebral cortex.

Above: The more frequently a synapse is activated, the more easily it will be triggered in the future, much like wearing a path through long grass.

70 HOW THE BRAIN MAKES MEMORIES

Fig. 5.1: Rat living in impoverished conditions

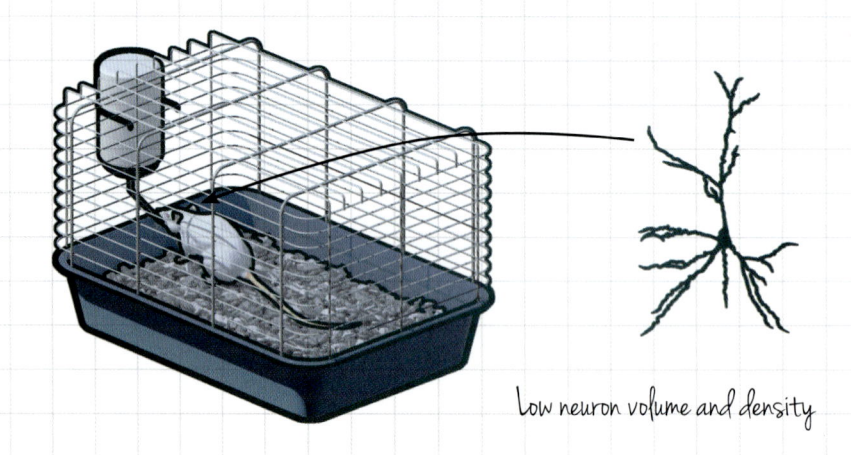

Low neuron volume and density

Rat living in rich conditions

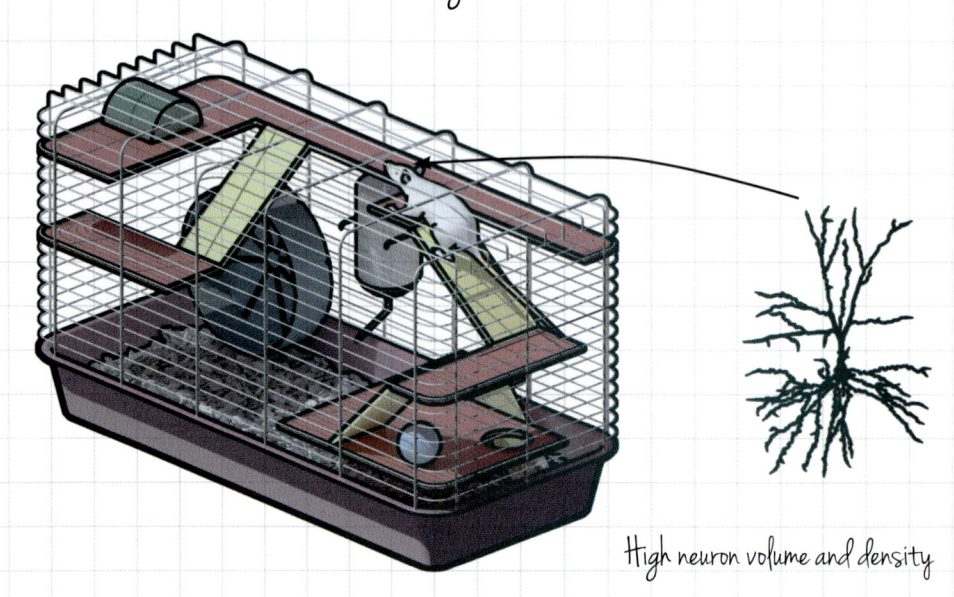

High neuron volume and density

biggest breakthroughs in our understanding of the brain. A particularly important follow-up study came from the American research psychologist Mark Rosenzweig and his colleagues in 1960. They took three groups of genetically identical rats and reared them under impoverished, simple and enriched conditions (*see* Fig 5.1, opposite). The rats in the impoverished conditions had fewer synapses, smaller brains and lower levels of the memory neurotransmitter, acetylcholine. In other words, the way the brains developed was hugely dependent on experience.

We now know that experience is constantly modifying our synapses and re-moulding our brain. Much of this understanding comes from research into the very simple sea slug, whose brain has large neurons and simple neural pathways, making it much easier to study than the more complex brain of a vertebrate. It is difficult to know how much of this detail translates to human beings, but there is good reason to suppose that many of the same principles apply. An important piece of research by Irish neuroscientist Eleanor Maguire has shown that the expert navigational abilities of London taxi drivers is strongly correlated to the amount of grey matter in their hippocampi. So, the fact that we can store memories and then refer to them later occurs because our brains physically change when we learn something new. This process is described as *neuroplasticity* and is discussed in more detail in Chapters 10 and 11.

Below: Research into the simple sea slug has been vital to our understanding of how experiences re-mould the brain.

THE ANATOMY OF MEMORY

So, we know that memories are stored across the cortex as neural networks and that these can be partly or wholly destroyed in patients who have significant damage to this part of the brain (for example, in Alzheimer's disease). However, this does not explain why the isolated removal of HM's hippocampus (*see* pages 65–6) should have caused him to experience such severe memory loss, nor does it explain why someone with damage to the cerebellum might no longer be able to play the piano. Biological explanations of memory also need to be able to account for phenomena that happen in people without any injury – for example, the complete amnesia we have for our infancy or the memory loss that happens in normal ageing. As you might expect, the anatomy of memory is quite complex and is distributed across many parts of the brain.

Let us start with the hippocampus. We saw earlier that significant damage to this area of the brain has a devastating effect on the ability of an individual to lay down episodic memories, and also that it is more developed in taxi drivers who are constantly exercising their route-learning memory. More recent studies have shown that the hippocampus is part of a bigger hippocampal formation, which includes other structures such as the *mammillary bodies* and the *fornix*. In fact, all of these parts were removed in the case of HM, and we now know that the whole system is important for remembering. The hippocampal formation is sometimes described as the "printing press", because its job is essentially to form memories and put them into storage. It is still not entirely clear how this happens, but it seems to be instrumental in forming associations – connecting and combining the cell assemblies.

Another area of the brain that works closely with the hippocampal formation is the prefrontal cortex. This is known to provide executive control and in many ways it acts like a librarian – organizing retrieval strategies, giving events a time stamp, verifying accuracy, deciding which memories should or should not be allowed into consciousness. Arguably, these are all the more "intelligent" aspects of remembering. The hippocampus is also very closely connected to the *amygdala*, an almond-shaped structure located deep in the medial temporal lobe of the brain, and this seems to provide the all-important emotional colouring for our memories. Patients who have damage to this area can recall events in rich detail, but they have no emotions attached to their memories, which can be very distressing.

An almost entirely separate part of the memory system lies in the cerebellum. This is where we store what people sometimes call *muscle memory* – a record of all the intricate

TRAIN YOUR BRAIN: HOW TO LEARN SOMETHING NEW

If you need to learn something – a script, a piece of music, a new move on the dance floor – there are lots of science-based tips to help you do so more effectively:
1. Make sure you are well rested: sleep is the time when the brain cements all of our memories;
2. Get moving: exercise increases circulation to the brain and may increase the number of synapses and cells we have in the hippocampus;
3. Pay attention: we learn better when we are giving it our full attention and are not distracted by anything;
4. Practice makes perfect: the more times we do, say or hear something, the stronger the connections in the brain will become. However, we remember more effectively if we have regular breaks;
5. Mix and match how you learn: the more varied you can make your learning, the richer the memory trace. So, for example, if you are learning a script, then read it, imagine it, listen to it, draw it, explain it – in fact, try as many different ways of looking at it as you possibly can.

HOW THE BRAIN MAKES MEMORIES   **73**

Fig. 5.2: Parts of the brain relevant to memory

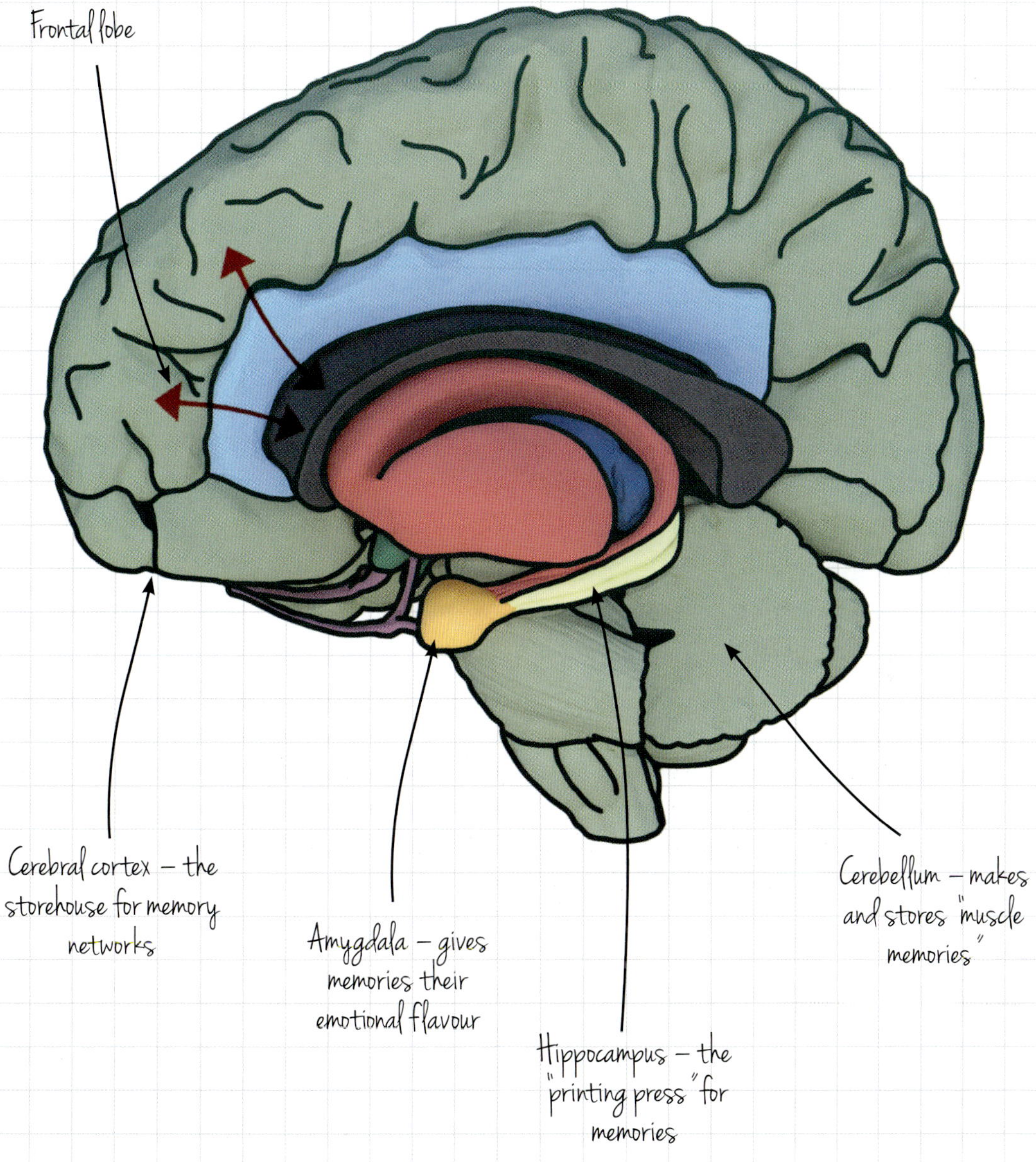

Frontal lobe

Cerebral cortex — the storehouse for memory networks

Amygdala — gives memories their emotional flavour

Hippocampus — the "printing press" for memories

Cerebellum — makes and stores "muscle memories"

Above: The muscle memory required for riding a bike is stored in the cerebellum.

combinations of movements that are needed for every simple or complex activity, from walking and balancing through to riding a bike, driving a car, playing football, or playing guitar. Again, it is possible to have isolated damage to this area, which will affect all of these skills but have no impact on remembering other things.

EVERYDAY MEMORY

Even normal memory can do strange things sometimes. Many of us have experienced *déjà vu* – the feeling of having experienced something before when, it is actually happening for the first time. When déjà vu occurs very frequently or in association with other symptoms such as hallucinations, it is usually a sign of underlying pathology in the brain (for example, epilepsy), or a psychiatric problem. However, research shows that up to 96 per cent of us will encounter it as part of our everyday experience. There have been various explanations for this mysterious phenomenon, but the most convincing one comes from work by the British cognitive neuropsychologists Martin Conway and Chris Moulin, at the University of Leeds. They suggest that normally the experience of remembering involves the reactivation of relevant cell networks that represent our initial experience but also additional activity that gives us the feelings of recollection. According to their theory, it is possible for cues in our environment to falsely activate the neurons that give us strong feelings of recollection, which leads to the unusual feeling that our present moment is actually a past one.

Memory may play tricks on us and is regularly fallible, but many experts believe that this is a natural and unavoidable consequence of its complexity and its necessary flexibility. As we have seen, because the memory system is distributed across large parts of the brain, it is quite vulnerable to damage.

However, this also means that it is almost impossible for all aspects of memory to be completely wiped out. Despite the devastating damage to Clive Wearing's memory (*see page 119*), he is still able to walk and play the piano. He also still knows who he is and where he went to university. Memories that seem to be particularly resilient are those that are automatic and unconscious. Old habits really do die hard, even in people with the most extreme memory loss, so making important things a habit is a good strategy for anyone with memory difficulties. We can support our memories in other ways, too – writing a diary can help to solidify experiences by reactivating those cell assemblies, especially if done just before sleep; boxes of memorabilia can provide excellent cues for reminiscence. Whatever age we are and however good our memory might be, we should never take it for granted. Our past experiences – conscious or conscious, actively recalled or popping unbidden into mind – will always shape our behaviour. Above all, memories are an intrinsic part of who we are.

Right: Writing a diary before bed helps to solidify experiences and supports good memory function.

MYTH BUSTER: MEMORY MISREPRESENTATIONS

In 2011, the American psychologists Daniel Simons and Christopher Chabris decided to examine the general population to test the notion that psychology is nothing more than common sense. They asked 1,838 US citizens a number of basic questions about memory and then compared their answer with those of experts. They found that a large proportion of the public hold false beliefs about the way that memory works, believing it to be more reliable and consistent than it really is. Here are some of the questions, along with the responses from the general public and experts:
(1) Do you think the testimony of one confident eye-witness should be enough to convict someone of a crime? *Public:* 37.1 per cent agreed; *Experts:* 16/16 disagreed;
(2) Do you believe the human brain works like a video camera, accurately recording what we see and hear so that we can inspect it later? *Public:* 63 per cent agreed; *Experts:* 16/16 disagreed;
(3) Do you believe that once you have experienced an event and formed a memory of it, that memory does not change? *Public:* 47.6 per cent agreed; *Experts:* 15/16 disagreed;
(4) Do you believe that hypnosis is useful in helping witnesses accurately recall details of crimes? *Public:* 55.4 per cent agreed; *Experts:* 14/16 disagreed.

6. ARE FEMALE BRAINS DIFFERENT FROM MALE BRAINS?

IS IT TRUE THAT MEN ARE MORE COMPETITIVE AND BETTER AT MATHEMATICS AND ENGINEERING? OR THAT WOMEN ARE MORE IN TOUCH WITH THEIR EMOTIONS AND BETTER AT LITERARY SUBJECTS?

From toys and toolboxes to tissues and tea, our supermarkets are littered with products that are presented and marketed differently for males and females. According to the campaign group Let Toys be Toys, gender stereotyping for children's merchandise has increased enormously since the 1970s. Neutral items such as colouring books or building blocks are branded as being "For boys" or "For girls", and identical chocolates and sweets are separately packaged in pink or blue. But do girls really prefer pink? Although some researchers have argued that there is an innate preference which reflects an evolutionary specialization for fruit gathering, other research has convincingly shown that it is a cultural phenomenon that starts to kick in around the age of two years old. Is the same true of toys and even careers? Is there any biological

ARE FEMALE BRAINS DIFFERENT FROM MALE BRAINS? 77

reason why girls might prefer dolls to cars, or be more likely to work in a caring profession? In his exceptionally popular 1992 book *Men Are From Mars, Women Are From Venus*, the American relationship counsellor John Gray states that love can only blossom between a man and woman when they learn to respect and accept their differences. So, to what extent do these differences really exist? And are they driven by nature or nurture? In this chapter we will take a balanced look at what science can tell us about brain differences in men and women, as well as examining how this relates to the way that people behave and think.

Opposite: Environmental and cultural influences on gender begin from the moment we're born.

Above: Toys that are specifically marketed for girls or boys can reinforce stereotyped beliefs about gender.

SEX DIFFERENCES IN THE BRAIN

Although the terms *gender* and *sex* are often used interchangeably, there is an important but tricky distinction between them. *Sex* has a clear biological definition and is determined by the physical characteristics that make a person male or female – their internal and external sex organs, hormones and chromosomes. *Gender*, on the other hand, is a less concrete term which essentially refers to the behavioural characteristics and roles that any given society identifies as being masculine or feminine. Some people make the assumption that this maps neatly onto *nature versus nurture* – in other words, any differences that are biologically driven are regarded as *sex differences*, and any that are culturally shaped can be described as *gender differences*. However, the categorization is not that simple. When we are examining any given psychological attribute – let us say emotional sensitivity or language ability – it is impossible to separate the contribution of nature and nurture. Cultural and environmental influences are so instinctive, deep-rooted and widespread that they will inevitably have an immediate impact, even for those infants whose parents try to maintain a gender-neutral approach in their upbringing.

This same problem arises when it comes to examining the brain itself. From the very moment our brains start to develop in the womb, the physical structure adapts in response to experience, as we will see in Chapter 10. Even a newborn baby will have been exposed to a unique set of chemical and physical events that shape which neural pathways are strengthened and weakened. Therefore, although we have the tools and methodology to carry out close examination of the structure and function of male and female brains, we can make relatively few assumptions about whether observed differences have a biological or cultural basis. There are some ways in which scientists have been able to tease these influences apart – for example, by looking at hormonal influences – but it is an ongoing and often controversial topic.

Above: Gender stereotypes are deep-rooted, which can sometimes make life challenging for those who do not conform.

VIVE LA DIFFÉRENCE!

So given this caveat, what if anything is actually different in the structure of male and female brains? During the 1980s and 1990s, a number of published studies suggested that men and women showed notable differences in the parts of their brains responsible for language. This assumption was partly based on reports that men were more likely than women to suffer language loss after damage to their left hemisphere, but it was also supported by a number of brain imaging studies. The overall suggestion was that the language functions were spread more evenly between the left and right hemispheres of women's brains than they were

Above: Early research suggested that women process language differently to men.

in those of men. American neurologist Norman Geschwind and his colleagues put forward a theory to explain this: they proposed that high levels of testosterone in early brain development slowed down growth of the left hemisphere in male babies, which led to it becoming more cramped. Although this theory was quite widely popularized and is still cited today, a number of follow-up studies found no difference in the size of left hemispheres in newborn babies of both sexes. In more recent times, Iris Sommer from Utrecht University has carried out a number of meta-analyses, which pool together the results from many relevant studies; these conclude that there is no evidence for these differences in the language regions of the brain.

One thing over which there is little argument is the size of the brain. It probably comes as no surprise to learn that men's brains are generally larger and heavier than women's – an average of 1,378g (3lb) compared with 1,248g (2¾lb). This, of course, is a direct reflection of the fact that the average man is taller and heavier (and by way of comparison, the brain of a blue whale is nearly five times bigger than a human brain), so in a way this does not tell us very much. What really matters is the size of different brain regions in relation to overall volume. Over the years, there have been many attempts to quantify this relationship, but the results are often contradictory. Recently, a team of researchers in Cambridge has gathered together all of the evidence from 1990 to 2013 in a scientific attempt to generate an overall picture of how male and female brains compare. Their report found that there are consistent sex-related differences in the volume and density of a number of areas, particularly those associated with memory, emotion and language – for example, the amygdala, hippocampus and areas of the frontal lobes.

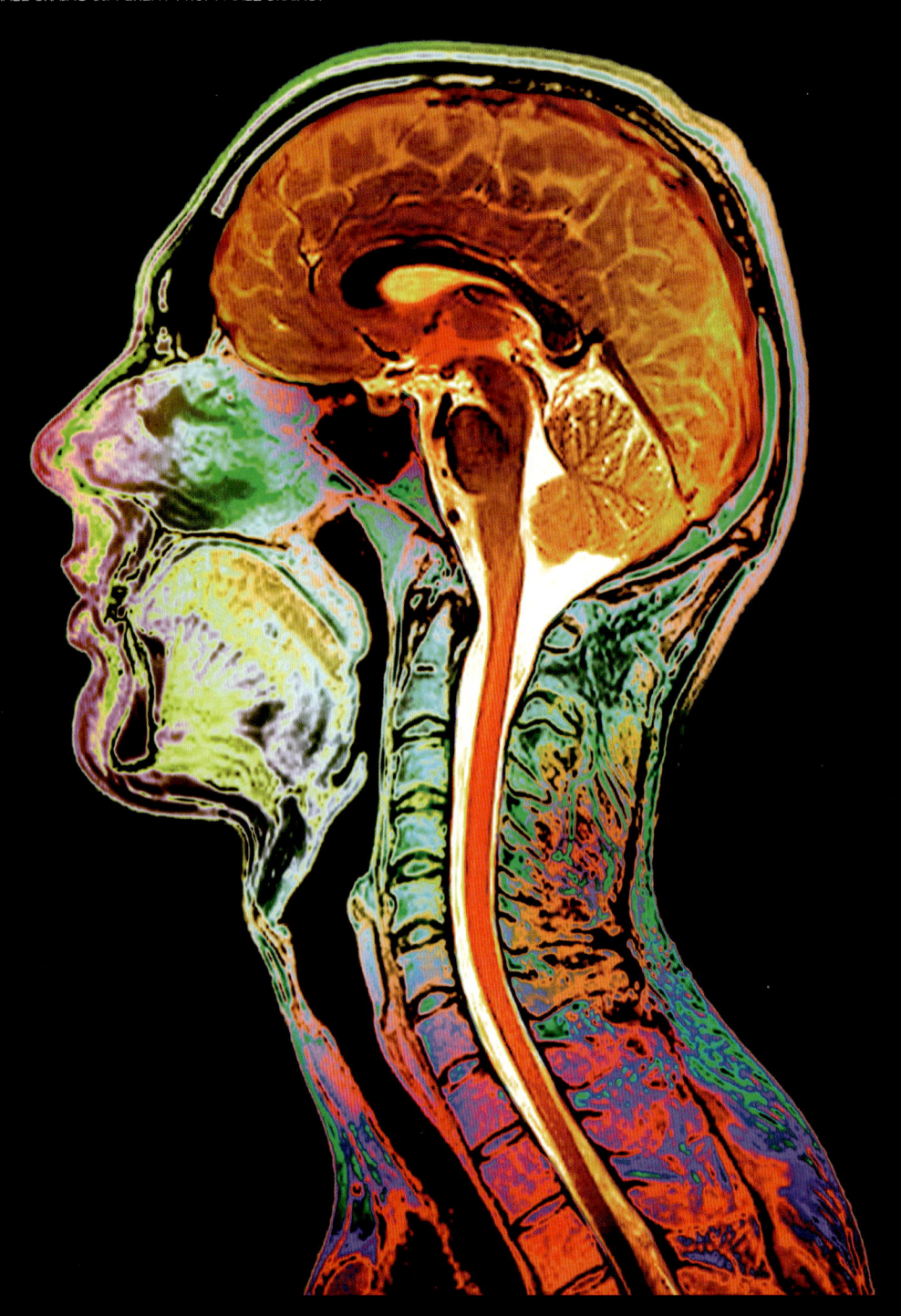

80 ARE FEMALE BRAINS DIFFERENT FROM MALE BRAINS?

There is a very important point to make here, though. What all of these studies rely on is a comparison between the *average* male brain and the *average* female brain. In reality, there is a massive overlap, meaning that any particular female may well have a more "male" brain than a particular man, and vice versa. Some very recent research led by professor of psychology Daphna Joel at Tel Aviv University specifically addressed this: by compiling brain-imaging data from 1,400 individuals, her team was able to produce a map of "femaleness" and "maleness" for each key brain structure. They then examined each individual brain to see how typical it was of their sex and found that fewer than eight per cent of brains fell into the category of being "all male" or "all female". Thus, although we can conclude that there are sex differences in brain structures, it is also true that very few people fall neatly into the "male" or "female" box – the vast majority of us lie somewhere on that continuum and some of us may well be positioned toward the opposite end for our sex. In fact, things are even a little more complex than this, because the differences can vary according to the area of the brain – for example, we may well have a very "male" amygdala and a more "female" hippocampus. According to Professor Joel, every brain is a unique mosaic, with each region having its own measure of "maleness" and "femaleness".

DO MEN AND WOMEN THINK DIFFERENTLY?

What, if anything, do these biological findings tell us about the way that men and women approach the world? We are led to believe that boys and men are better at spatial reasoning and logical academic subjects, such as mathematics, as well as being more aggressive and impulsive and less emotionally

Opposite: Consistent but small differences are found between an average female and male brain.

Below: Are men better at some tasks and women better at others?

sensitive (*see* MYTH BUSTER box, page 87). Likewise, women are seen to be more suited to linguistic tasks, better at reading emotional cues and more able to multi-task. In the last 40–50 years, gender stereotypes have been increasingly challenged and yet we still live in a world where less than a quarter of the world's managers are female, the gender pay gap in the UK is around 20 per cent and only 1 per cent of the titled land on Earth is owned by women. This is despite the fact that girls consistently outperform boys when it comes to final secondary school exams. Many cultures also place huge demands on women in terms of physical appearance, a significant contributory factor to an increase in the levels of eating disorders such as anorexia nervosa. A recent survey by the organization Girlguiding UK found that 87 per cent of females between the ages of 7 and 21 perceived that they were judged more by their physical attractiveness than their ability. Stereotyping can be equally harmful to men, with young boys feeling under pressure to conform to expectations of being rough and tough. While female suicide rates have halved in the last twenty years, male suicides continue to rise. This phenomenon is largely attributed to increased pressures to be the breadwinner in an increasingly competitive financial environment, alongside a taboo when it comes to discussing emotional problems – a "big boys don't cry" mentality.

These are worrying statistics, regardless of whether they are biologically or culturally determined, so it is important

Below: Females outperform males in musical achievements at school level, yet professional orchestras typically comprise around three times more men than women.

Above: A meta-analysis by educational psychologist Erin Pahlke showed that neither boys nor girls benefit from single sex education.

to use science to examine and challenge this tendency to pigeonhole people. Some things are not in dispute – we know that there are obvious biological distinctions between men and women in terms of sex organs; there is a little more overlap when it comes to physical attributes such as height and strength but, as we have seen, even less discrepancy when it comes to brains. Additionally, as discussed earlier, any brain differences may simply reflect individual experience – for example, the developmental trajectory of the emotional regions of a young boy's brain will be hugely dependent on the way his parents and others behave toward him in emotional situations. The same is true when it comes to psychological and cognitive skills. For example, there is consistent evidence of a small sex difference in a number of tasks; on average, men outperform women in particular spatial tasks, especially the ability to mentally rotate an object in their imagination, and women seem to be better at particular language and memory tasks. However, once again these differences are generally small, show significant overlap and are also heavily influenced by experience and expectation.

In her book *Delusions of Gender*, the Canadian-British psychologist Cordelia Fine challenges some of the scientific evidence for cognitive differences between men and women. She argues that prior opinions will affect the way an individual approaches and completes a task. So, for example, if a woman has an underlying belief that females are better at language tasks, then their confidence and competence will rise for any task that requires this skill. To put this into more everyday terms, a man becomes better at parking his car simply because that is what he expects of himself in light of common stereotypes. This expectation effect is commonly found throughout many psychology experiments. Cordelia Fine also points to more global evidence that the percentage of women in science-based jobs is higher in countries that conform less strongly to the image of male scientists. To illustrate this specifically, Diane Halpern, an American psychologist who specializes in this field, cites a piece of research in which the same spatial test was given to boys and girls: when it was described as a test of "geometric ability" the boys did better, whereas when it was called a test of "drawing ability" the girls came out on top. Nevertheless, Halpern does conclude that there are some genuine but small sex differences in some cognitive tasks.

Christian Jarrett, author of *Great Myths of the Brain*, says that one of the biggest traps we can fall into is an assumption that, where there are sex-related brain differences, these will necessarily lead to an effect on intellect and personality. This is a mistake made by John Gray, the proponent that "men are from Mars" and "women are from Venus". He refers to early neuroscientific evidence that females tend to use both brain hemispheres for a task, while men are more likely to have focused activity on one side. Gray extrapolated that men could therefore only think of one thing at a time while women could easily multi-task. Not only has this brain evidence now been dismissed by a number of other researchers, but also this is a huge and false leap to make. Gray is not alone – since 2002, Leonard Sax has been campaigning for single sex education in the USA on the basis of brain differences between boys and girls. His reasoning is based heavily on a single piece of research that suggested a delay for boys in the connections between their amygdala and cerebral cortex, which he suggests will affect their emotional functioning. This was an interesting but small study with many flaws, and in fact a big meta-analysis in 2014 showed that single sex education confers no advantage on either boys or girls.

SEX HORMONES, THE MALE BRAIN AND AUTISM

Despite the fact that myths and misunderstandings abound, Jarrett and others stress that we should not throw the baby out with the bathwater – it is important to acknowledge that some differences do exist, because these have scientific worth. In particular, research on sex differences may shed light on

Above: John Nash suffered from schizophrenia, a condition that generally occurs earlier and possibly more frequently in men.

a number of neurodevelopmental conditions that have a male bias – for example, autism, attention deficit disorder, Tourette's syndrome and schizophrenia, as well as those in which women are disproportionately affected, including Alzheimer's disease and recovery from brain injury. Simon Baron-Cohen, Professor of developmental psychology at the University of Cambridge, has spent most of his career trying to understand the biology and psychology of Autistic Spectrum Disorders. Classic autism affects approximately four times as many boys as girls, while Asperger's syndrome may occur fifteen times more often in males than females. Baron-Cohen argues that although social variables are extremely important, biology must and does play a role in gender differences, even if it is only a small one. If we ignore this premise, he says, then we are going back to the days of believing that autism is caused by familial or social factors, an idea that science now strongly refutes.

Baron-Cohen puts forward a hypothesis that autism occurs in individuals who have an extreme form of the male brain. In this theory, he acknowledges that there is an enormous crossover between the typical male and female brain, but nevertheless argues that biological differences do exist and are already in place by the time of birth. One way he tested this was to examine the behaviour of newborn infants when they were presented with either a face or a mechanical mobile. His team measured the amount of time the babies looked at either of these stimuli, making sure that there were no clues about the sex of the infant. They found that on average the boys spent longer looking at the mechanical mobile, while the girls preferred the faces. (This finding is in keeping with some parents' beliefs that there seem to be innate preferences dictating which toys most attract their sons or daughters.) Baron-Cohen also reports sex differences in many other behaviours, although once again we must bear in mind that statistical findings always rely on averages and that any child that is more than a day old will already have been exposed to the environmental influences of stereotypes.

The idea that autistic spectrum disorders may reflect an extreme end of the male spectrum is not new – in fact, it was one of the suggestions put forward in the 1940s by the Austrian paediatrician Hans Asperger, one of the first people to characterize and define autism. In line with Geschwind's theories from the 1980s, Baron-Cohen's proposal relies on the idea that early brain organization is heavily shaped by levels of prenatal hormones. He points to the fact that left-handedness is more common in boys (twelve per cent) than in girls (eight per cent) and also to evidence that measures of prenatal testosterone taken from babies in the uterus

Below: Prenatal levels of testosterone and oestrogen influence the organization of the brain.

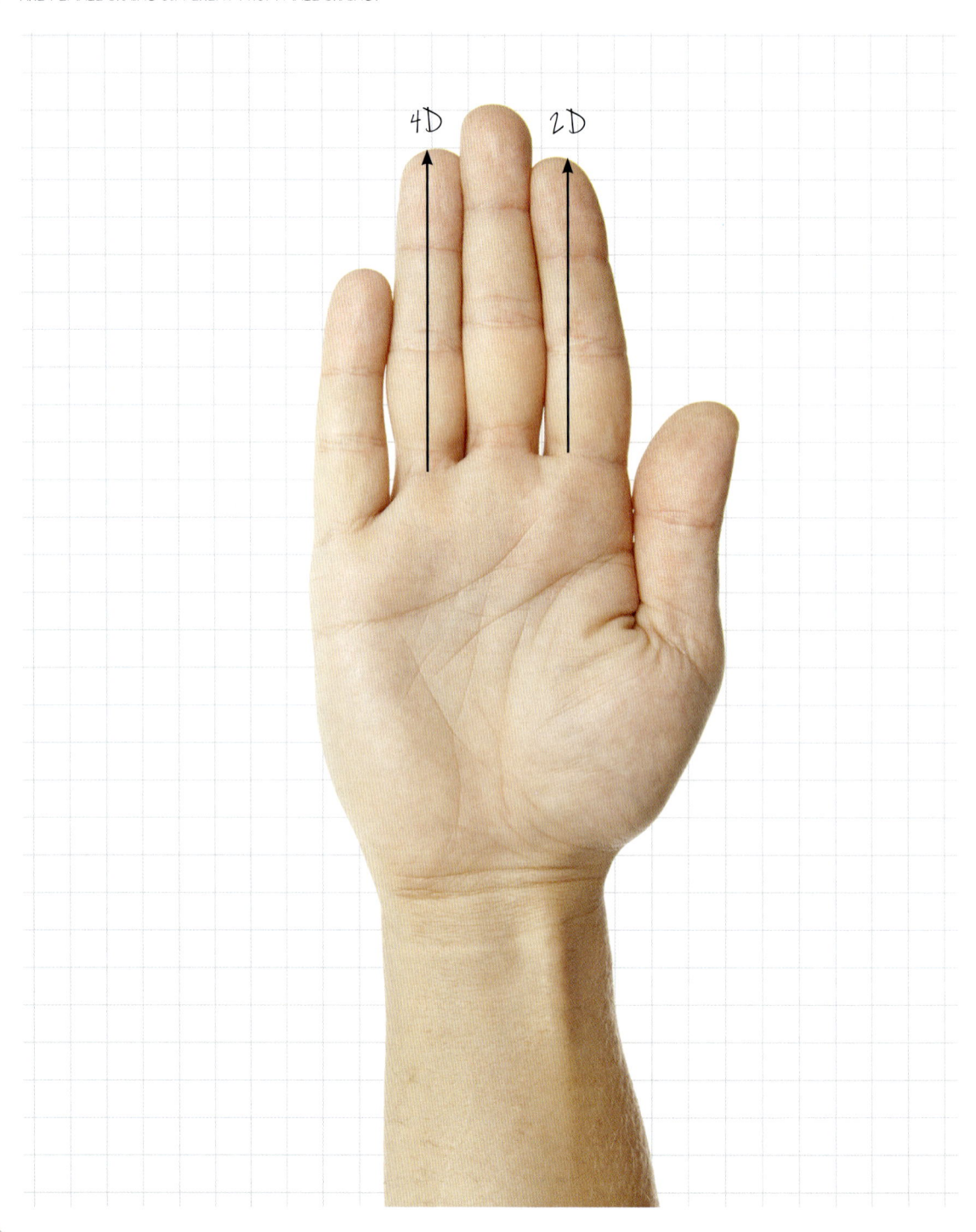

86 ARE FEMALE BRAINS DIFFERENT FROM MALE BRAINS?

consistently correlate with later behavioural development. One fascinating line of research has looked at ways of estimating prenatal hormone exposure in fully-grown adults. This work by John Manning, a psychologist from Swansea University in Wales, suggests that the ratio of our second to fourth finger is a good proxy for the balance of oestrogen and testosterone that we were exposed to in the womb: the greater the length of the fourth finger in relation to the second finger, the higher the proportion of prenatal testosterone. Although this sounds like a strange idea, this ratio does seem to predict cognitive skills, career choice and even fertility, so there is some support for it.

This chapter has focused fairly heavily on brain differences and how they might relate to behaviour, but as we have just seen, hormones also play a part. As well as their possible influence on brain development, there is good evidence that levels of circulating testosterone and oestrogen affect spatial and verbal skills, the ability to detect emotional expressions, face recognition, aggression, partner preference, sporting performance and sex drive,

among others. What is interesting about this research is that it suggests some of the observed sex differences may relate less to structural brain differences and more to the levels of a given hormone at any one time – this can affect women over the course of their menstrual cycle, during pregnancy or at menopause, or both men and women who undergo any kind of hormone treatment. Again, though, it is important to be aware that findings are not always consistent, and one problem with all research on sex differences is that there may be a publishing bias – a tendency to publish work that is exciting and supports people's intuitive ideas, while relegating other findings to the bottom of the filing cabinet. Having said that, it is fairly safe to say that there are differences in the *average* male and female brain, but also important to emphasize that many people fall well outside these parameters. There is scientific merit in recognizing and understanding the effects of genetic and hormonal factors on the brain and behaviour. However, it is also vitally important that we maintain a high awareness of the influence of society and expectation, not just on how we behave or what career we choose, but on the development of the brain itself.

Opposite: Scientists believe that we can estimate levels of prenatal testosterone by looking at the ratio between the second and fourth finger.

MYTH BUSTER: BOYS ARE BETTER AT MATHS THAN GIRLS AND MEN ARE MORE AGGRESSIVE THAN WOMEN

Whether you are reading a magazine, surfing the Internet, or engaged in a conversation with a friend or work colleague, you will probably come across comments about sex differences on a daily basis. Often these are backed up by "factual" pieces of science, usually relating to the biology of male and female brains. With a bit of research you can probably debunk many of the myths yourself, but let us look briefly at two here:

1. *Boys are better at maths than girls*: although there are a few more boys achieving the top one per cent of maths grades, statistics from across the globe show that there is no difference in the average maths scores of girls and boys at the end of their schooling;

2. *Men are more aggressive than women*: One very neat study showed that women were indeed less aggressive than men on a violent computer game while they were playing in the same room. However, as soon as they were allowed to play on their own, the women actually dropped more bombs than the men and showed greater hostility to other online players.

7. WHY DO WE CRY AT FILMS AND LAUGH AT JOKES?

OUR RICH PALETTE OF EMOTIONS IS A DEFINING AND ESSENTIAL FEATURE OF HUMANITY - BUT WHAT ARE EMOTIONS ACTUALLY FOR?

In the *Star Trek* film *Generations*, the quirky but strangely likeable android character named Data is given something rather special – an emotion chip. This fictional piece of technology is a major step forward in Data's lifelong quest to understand what it is like to be human and his story provides us with an interesting thought experiment. What would it be like to be devoid of any emotions or to be able to switch them off at will? Do we need to have feelings in order to understand each other? Do emotions help or hinder our ability to function in this world? Would it matter if we never felt sadness, happiness, desire, jealousy, rage, pride, fear, gratitude or love?

THE IMPORTANCE OF EMOTION

Although the word *emotion* did not enter the English language until the mid-sixteenth century, theories about the nature and role of emotion go back to the ancient Greek philosophers Plato and Aristotle, who highlighted the reciprocal connections between cognition (thinking) and emotion (feeling). Modern science supports this basic idea and there is now substantial evidence that emotions play a vital role in our judgement and decision-making. Sometimes this is at a conscious level – for example, we may choose to spend time engaged in a particular activity because we remember that we previously enjoyed it or that it made us feel good. Likewise, we may avoid certain situations because they remind us of an earlier experience that left us feeling uncomfortable or unhappy. However, emotions can also affect our decision-making at a more unconscious level. Many people talk about making a choice because of a "gut reaction", or may agonize about whether to follow their "heart" or their "head" when they are faced with a difficult dilemma.

Often these more intuitive decisions can turn out to be the right ones, despite the fact that they may seem to go against logical reasoning. So does this common experience suggest that our body may sometimes know best? If so, why would

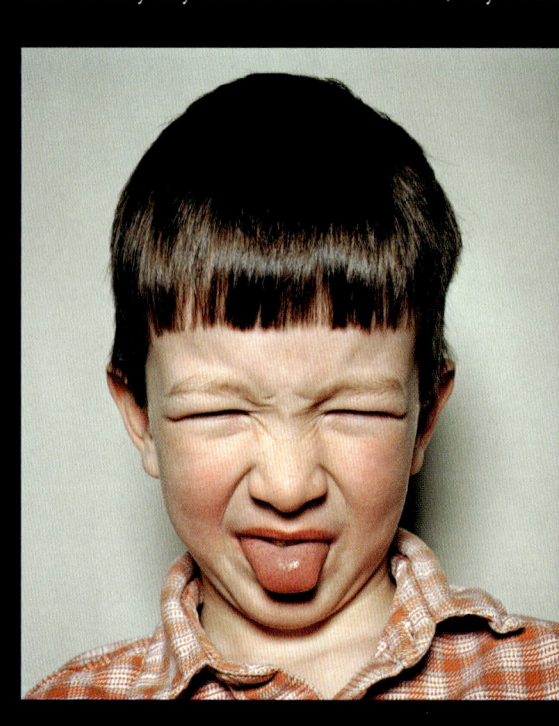

Above: Our ability to recognize and describe "disgust" begins at around the age of five.

Above: Emotions are central to our relationships and play an important part in bonding and attachment.

this be? Antonio and Hanna Damasio, neurologists who make a husband and wife team, have carried out a series of important studies, showing that decisions are based not just on rational thought but also on subtle physiological signals, such as a change in heart rate or an increase in sweat. Their *somatic marker hypothesis* claims that these are essentially emotional signals that act as an unconscious reminder of the outcomes of previous similar experiences. This system does not replace the need to think carefully about decisions because every situation is unique, but it does provide us with additional important data. These somatic markers have been shown to be particularly helpful when we need to make decisions under pressure.

Emotions do not just guide our thinking, they are also central to our relationships and play an inherent part in our bonding and attachment to one another. So, while we rely on our feelings to direct us in deciding with whom we might like to spend time, we also use emotions such as laughter and crying as a powerful way of connecting and communicating. This is one reason why live comedy nights, musical concerts and sad or exciting films can be good for nurturing friendships. According to Sophie Scott, a neuroscientist from University College London, we are 30 times more likely to laugh if we are with other people than when we are on our own, and the more we know and like someone, the more likely we are to laugh. Shared laughter

normally be portrayed through gesture and tone of voice, but which can easily get lost in written communication, particularly when it is condensed (as when texting). The huge and rising popularity of emojis, first developed in Japan in 1999, is just one interesting sign of the human desire to share emotion.

DEFINING EMOTIONS

Joseph LeDoux, a world-leading expert on the neuroscience of emotion, says that *one of the most significant things said about emotion may be that everyone knows what it is until they are asked to define it.* It is true that it is difficult to describe exactly what an emotion is, despite the fact that it is an overwhelming aspect of our existence. Dictionary definitions vary, but they all tend to refer to "feelings", which highlights the very physical nature of emotive experience. Think back to the last time you felt really excited about something. Can you remember how it actually felt? If you can really place yourself in the moment, you might even be able to recreate some of the experience now – a slightly racing heart, maybe a little bit of tension in the muscles, possibly a small smile on your face. Now imagine the same experience without any of those physical sensations: you will probably find that it is quite difficult to conjure up any real sense of excitement without them.

Many of the physical qualities of excitement and other emotions comes from activation of the Autonomic Nervous System, the fight-or-flight mechanism that alters heart rate, breathing patterns, saliva production, contractions in the gut and sweating, among other things (*see* Chapters 1 and 4 for more detail). However, emotions can have other very physical manifestations – laughter, crying or screaming, for example, and of course facial expressions and non-verbal signals, like body posture or hand gestures. Together, these observable characteristics of emotion are referred to as *emotional expression* and they can be measured and identified fairly readily in both humans and animals.

As well as being a useful tool for psychological research, there are legal and commercial motivations for assessing emotional state, which have led to the development of various technologies that purport to read someone's mind. For example, *lie detectors* have existed in one form or another since the early 1900s: these devices essentially measure the physiological signs associated with the anxiety and discomfort that someone experiences when they lie. Although many people had high hopes that this might transform the justice system, lie detectors actually remain a relatively unreliable technology. More recently, there has been an explosion in the commercial development of emotion-recognition technology for other purposes. For example, in December 2015, research showed that the way a person moves their computer mouse reveals their level of frustration. This could provide important data on the usability of websites and software but also has the potential to feed into online social interactions.

On the whole, humans are generally far more adept than machines when it comes to reading and interpreting emotional expression. This essential skill is the means by which we establish the likely intentions or motivations of our fellow human beings – at a very basic level, whether they are likely to be friend or foe. However, this window into another person's emotional state is also a very important aspect of understanding and responding appropriately to those around us. It is at the very heart of what allows us to connect with each other, and those who have difficulties

Above: The popular trend of using emojis provides a fun way for people to share emotions in online communication.

reading emotion – such as individuals with an Autistic Spectrum Disorder – can really struggle to form relationships and may find the world a bewildering place.

From a scientific perspective, being able to measure emotional expression has been an important part of developing our understanding, but in reality this is only half the story. Anxiety and anger, for example, lead to quite similar bodily responses – increased heart and breathing rate, muscle tension, sweating – and yet they are quite different subjective experiences. Thus, psychologists make a clear distinction between emotional expression and *emotional experience* – the conscious mental state associated with an emotion. Although this can be inferred through observation, it is only truly known by the individual experiencing it. With human participants we have the option of asking people to describe how they are feeling, but this is a significant limitation in animal research.

EXPLAINING EMOTION

The nineteenth-century evolutionist Charles Darwin was one of the first people to investigate emotion scientifically, but the earliest established psychological theory came from William James and Carl Lange, who independently put forward a similar explanation in the late nineteenth century. Their idea, now known as the James-Lange Theory, was that the physical sensations come first; in other words, we have automatic physiological responses to different situations and this results in an emotional feeling. James and Lange believed that we become scared because our heart is beating faster or that we are happy because we laugh. There is certainly some support for this: for example, people will find cartoons and jokes much funnier if they are asked to put a pencil horizontally between their teeth while they watch them (try it – it forces your mouth into a smile). Yet the general consensus is that this explanation is too simplistic and cannot account for the full breadth of human emotional experience.

As a direct test of this theory, the American physiologist Walter Cannon carried out a series of experiments in the 1920s, in which he cut specific nerves in cats to prevent normal automatic bodily responses to a fearful situation. He found that the cats continued to display fear and aggression,

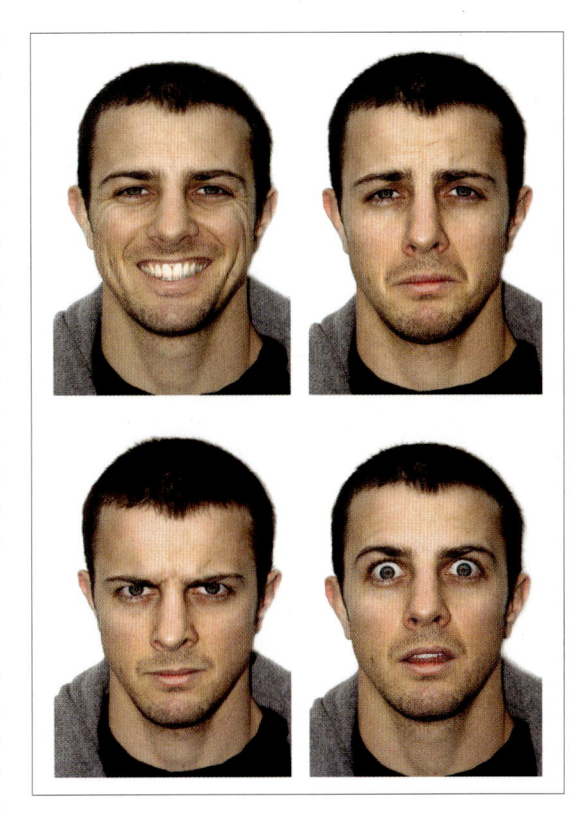

Above: Paul Ekman found six core emotions that are universally understood. Can you identify anger, happiness, fear and sadness?

and along with his student Philip Bard, he presented a new account – the Cannon-Bard Theory of emotion. They argued that the physical response and the subjective experience are two distinct aspects of emotion, which can occur independently and in any order: our bodies have a fight-or-flight response and at the same time we perceive that we are in a frightening situation. Together these give us a feeling of fear. Again, there is some support for this theory: we know that people who lose sensation in their body as a result of spinal injuries may sometimes encounter less intense feelings of anger and excitement, but they are certainly not devoid of emotion.

A very good compromise between these two approaches was proposed in 1962 by yet another duo, Stanley Schachter and Jerome Singer. They suggested that we have automatic

Above: Emotional expressions refers to behavioural manifestations of an emotional state.

physical responses to particular situations, but then use an appraisal of our current circumstances to interpret which emotion we might be feeling. In a fascinating experiment, they pretended to test a new drug on different groups of people, but in fact injected them with adrenaline. Some participants were told what side effects to expect and others were not, but they were all asked to wait in a room with the experimenter's accomplice, who was behaving in either an angry or euphoric way. They found that those people who knew what physical symptoms to expect – a pounding heart, flushed face and quicker breathing – did not report high levels of emotion. However, those who had not been given this information described themselves as feeling very agitated or excited, depending on whether they had been partnered with the angry or euphoric person. The important thing to note here is that individuals interpreted the sensations as emotion only when they did not know to expect them, but crucially the specific label they gave to their emotion was determined by the behaviour of the people around them.

EMOTION IN THE BRAIN

Although there have been more recent theorists – Damasio and LeDoux in particular (*see* pages 89–90) – the Schachter-Singer model of emotion still sits very well with modern neuroscientific accounts. People investigating the brain basis of emotion can be roughly divided into two camps: those who believe that some primary emotions consistently and specifically correspond with identifiable brain regions; and those who believe that emotional experience results from activation across more general brain networks. Despite this disagreement, there is a good consensus about which brain structures are important; these include the amygdala, hypothalamus, hippocampus, anterior cingulate, insula, basal ganglia, prefrontal cortex and cerebellum (*see* Fig 7.2).

Early accounts of emotion in the brain referred to what we now call the *limbic system* – a complex network of structures embedded deep in the organ with rich connections to the prefrontal cortex. Although this system was first described by the nineteenth century French physician Paul Broca (*see* page 16), it was not until 1932 that the American neuroanatomist James Papez identified it as the site of emotion. While we know that there are other important brain regions, there is now substantial evidence that the key elements of the limbic system are vital for emotional perception and response. For example, the hypothalamus controls the Autonomic Nervous System, which produces many of the important physiological effects described earlier. It also controls the production of testosterone, which plays an important part in aggression and sexual motivation. Early experiments showed that removing different parts of the hypothalamus in dogs led to changes in their emotional displays, and there are many examples of how this hormone affects human behaviour. For example, footballers show higher levels of testosterone when playing a match at their home ground, which boosts aggression and may account for an increased chance of winning those games.

There has also been a huge amount of work on the amygdala – a brain structure that has been linked with everything from antisocial behaviour in children to political ideology and voting intentions. Fascination with

WHY DO WE CRY AT FILMS AND LAUGH AT JOKES? 93

Fig. 7.1: Theories of emotion

94 WHY DO WE CRY AT FILMS AND LAUGH AT JOKES?

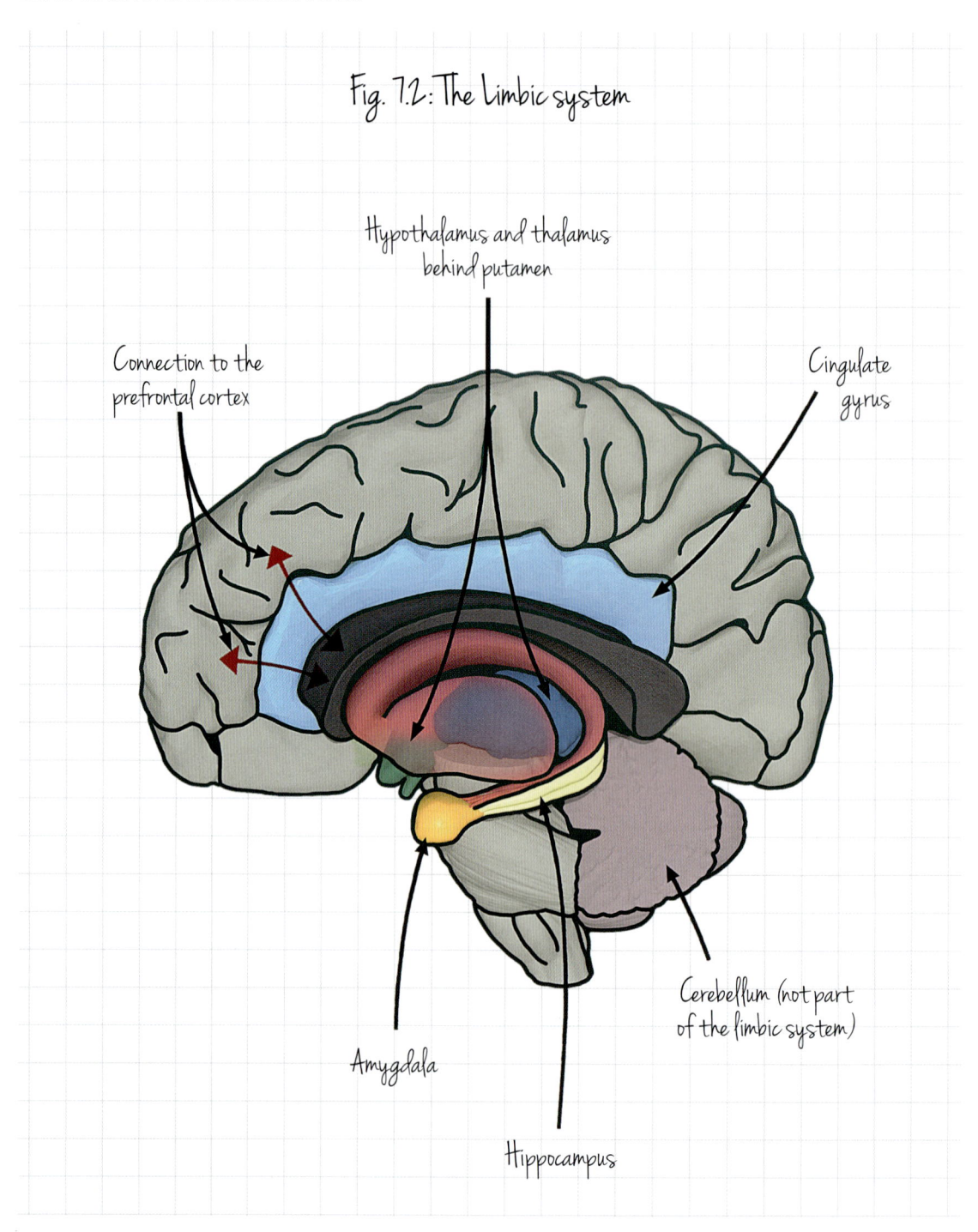

Fig. 7.2: The Limbic system

Hypothalamus and thalamus behind putamen

Connection to the prefrontal cortex

Cingulate gyrus

Amygdala

Hippocampus

Cerebellum (not part of the limbic system)

WHY DO WE CRY AT FILMS AND LAUGH AT JOKES? 95

the amygdala began with a series of studies in the 1950s and 60s in which the amygdaloid nucleus and surrounding structures were removed from monkeys and other animals. This caused quite significant but varied emotional changes – for example, reduced fear, less aggression and changes in sexual behaviour. It even affected which monkey was "in charge" of a colony. When this region of the brain is naturally damaged in humans, they seem to experience less fear but are also less able to recognize when other people are feeling scared. Artificially stimulating the amygdala can cause heightened fear responses, and there seems to be increased activity in some people with anxiety disorders.

The idea that fear resides in the amygdala has been used to support the "locationist" view of emotion – that identifiable emotions map directly onto specific structures. Other work has shown that activity in the insula signals disgust and that the prefrontal cortex helps to manage, colour and control our emotions (*see* Case Study, page 23, Chapter 1). However, in many ways we have come back full circle to the original ideas of James Papez, with most neuroscientists agreeing that emotions are best represented by the combined activity of many different structures within the brain.

Recent work from Kristen Lindquist at Harvard University pulled together the brain imaging data from numerous different studies and proposed a theory which was very much in line with the findings of Schachter and Singer. Their model suggests that physiological effects occur in response to stimuli, often those that relate in some way to previous experience, and then we use higher cognitive functions to make meaningful connections between those sensations and our current surroundings. The first part of this process is driven by the limbic system and surrounding structures, while the appraisal and control part seems to require input from the outer layers of the brain, in particular the prefrontal cortex. Overall, this system enables us to recognize sensations, such as a pumping heart or a flushed face, and then determine whether they might be a physical symptom – resulting, for example, from illness or having just run up two flights of stairs – or instead a discrete emotion, such as excitement or love.

Top: Surgical removal of part of the hypothalamus in dogs leads to displays of rage.

Above: Surgical removal of the amygdala in monkeys causes them to relinquish leadership.

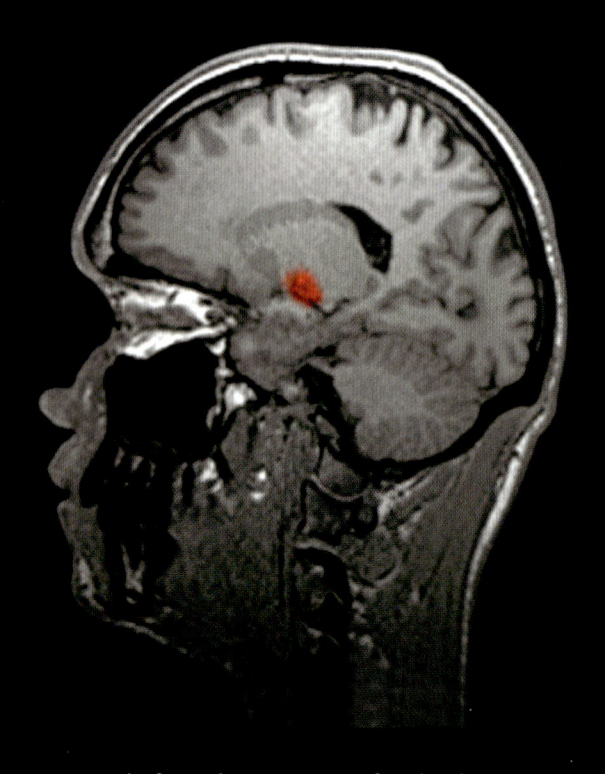

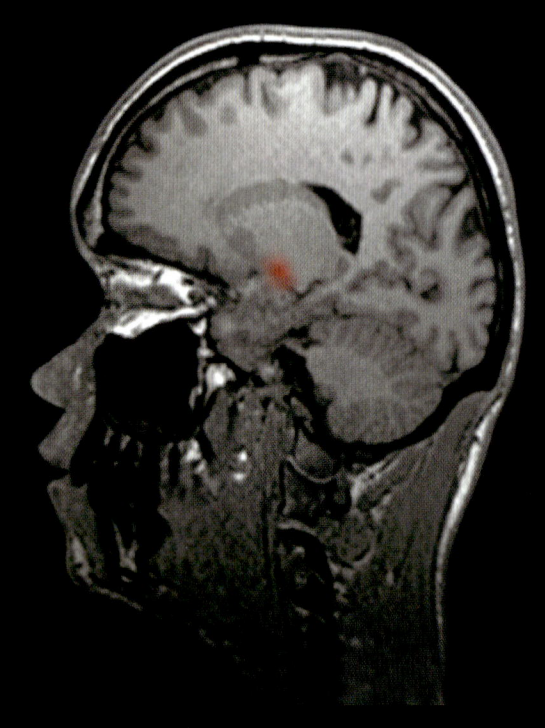

Above: The amygdala has been linked with many emotionally driven behaviours, from antisocial behaviour to voting intentions.

Research from the University of Colorado in 2015 consolidated the locationist and constructionist views described on page 95. Using complex mathematics to analyze brain scans, they confirmed that each of the basic emotions – fear, anger, disgust, sadness and happiness – involved a network of different brain structures, but the researchers also demonstrated that different emotions evoked distinct reliable patterns of activation across multiple brain networks. In other words, they showed that none of these emotions resides in one region, but they are nevertheless unique in terms of their overall brain activity.

DISRUPTED EMOTIONS

Most brain research has focused on a relatively small number of core emotions, but the big remaining question is how we are able to experience and identify such a huge breadth of different emotions. There is obviously a developmental and cultural element to this – we have to learn to recognize particular feelings and be able to use our experience and evaluative skills to identify what they represent so that we can act accordingly. Can you identify how you are feeling right at this moment? What physical clues are you using? Can you describe it? Do you know why you are feeling as you do? Being able to recognize and label emotions seems to be an important factor in terms of everyday functioning and wellbeing, and is something that some people are better at than others. *Alexithymia* is a personality trait in which people find it pathologically difficult to identify and describe their own emotional states. Although it lies on a continuum, it occurs at a clinical level in about ten per cent of people and is particularly prevalent in individuals with anorexia nervosa, depression, panic disorders and autism. Alexithymic tendencies can lead to huge difficulties with interpersonal relationships and social functioning.

Emotions can be disrupted in many other ways – anxiety disorders, depression, insecure attachment, difficulties regulating emotions, excessive euphoria – often to the extent that an individual is unable to function normally in the world. Sometimes there are obvious organic causes, such as in the case of an individual who has suffered a brain injury or has dementia, or someone under the influence of psychoactive drugs. But often, emotional extremes are the result of an experience (for example, bereavement or post-traumatic stress disorder) or a whole lifetime of challenges. On other occasions, there is simply no obvious cause, which makes it particularly difficult for the individuals concerned and those around them. As we saw in Chapter 3, medication can sometimes help those who are plagued by difficult emotions, but psychotherapies can also be very effective. These often target the processes described above, by teaching individuals how to recognize and control the physical sensations, and/or helping them to change the way in which they interpret, evaluate or manage them.

An intense bout of depression or anxiety can be utterly debilitating, but equally a period of complete emotional numbness leaves people unable to make decisions, form attachments or have any drive to complete even basic everyday tasks. Neuroscience has shown us that, unlike in *Star Trek*, there is not a single, distinct emotion centre in the brain that we can switch on and off. Yet however difficult they might sometimes be, it is emotions that give our world colour – and they are an absolutely crucial part of our existence. Uncomfortable and painful feelings have a vital role in teaching us what or who to avoid in future and they also provide context for the times of joy and happiness which propel us forward in life. From time to time a situation can be so intense that it evokes apparently conflicting but somehow complementary emotional responses. *I am happy and yet I am crying!* said *Star Trek's* Data, when he found his cat. *Perhaps the chip is malfunctioning?!* His shipmate, Troi, smiled at him. *I think it is working perfectly*, he said.

Above: Positive emotions provide motivation and propel us forward in life.

CASE STUDY: MB

Trudi Edginton and Ashok Jansari, British neuropsychologists, report the case of a 34-year-old woman, whom they call MB. After suffering from a seizure, MB was found to have a small tumour in her right temporal lobe. Surgeons were able to remove the tumour, which turned out to be benign, but this also meant taking out some of the surrounding brain tissue. MB recovered well but was left with an unusual impairment – although she could remember events perfectly well, her memories had lost of all their emotional content. So, while she could describe her wedding day in some detail and had no problem recalling the moment when she heard about the twin towers falling on 9/11, she did not have any feelings at all associated with these moments. Interestingly, MB was still able to identify emotional expressions in other people's faces and voices, so this deficit seemed to be specific to memory. This appears to suggest that the emotional colouring of memories has its own distinct circuit, which was damaged in this patient.

8. HOW DOES THE BRAIN TELL THE TIME?

TIME, ACCORDING TO DR WHO, IS "LIKE A BIG BALL OF WIBBLY-WOBBLY, TIMEY-WIMEY STUFF."

This definition, dreamed up by Steven Moffat, one of the writers of the BBC science-fiction television programme, is in many ways as good a description as any other. Time is a mystical, elusive, intangible and arguably subjective concept. As the great Albert Einstein himself said, time is an illusion, and the only reason for it is that it stops everything happening at once. It is certainly a phenomenon that has puzzled philosophers, physicists and neuroscientists alike, although one thing most scholars agree on is that time is a human construction. It is our attempt to quantify and measure the changes that happen as light becomes dark, winter turns to summer, young turns to old, or one moment becomes the next. Time provides a vital thread through our lives that allows us to predict and coordinate our actions, and gives context and meaning to everything we do, think and feel.

UNDERSTANDING TIME

Our sense of time passing is intriguing. It can mysteriously stretch and shrink depending on how bored we feel, how busy we are, how much sleep we have had, whether we have

Above: Our fascination with time goes back to at least 2000 BC.

Right: Our sense of time passing can mysteriously shrink and grow depending on what we are doing.

Above: We respond faster to a bang than a flash, which is why races are often started with a gun.

had a glass of wine and how old we are. Even when we wake up from sleep we usually have a good sense of the amount of time that has passed, although interestingly this is not the case when there is a complete loss of consciousness (for example, when someone is under general anaesthetic). Overall, humans are incredibly good at judging time, particularly at small intervals – production and understanding of speech, for example, is critically dependent on our micro-timing skills. Time perception also seems to be a very robust skill: while head injuries or strokes can severely impair important functions such as language or memory, there are few patients who lose all of their timing abilities.

Our ability to place past or present events into a time framework not only provides us with a narrative – it is crucial for survival. The brain's timing mechanisms enable us to perform lots of everyday behaviours, from predicting the arrival of food and coordinating the necessary actions, through to estimating when it is safe to cross the road. We also use time judgments for many perceptual functions – for example, locating sounds in space; if one voice is played simultaneously through two speakers it will sound like a single person, but separate them by just 50 milliseconds and they become two distinct voices. Bats, and even humans to a lesser extent, can estimate the size and shape of a space they are in, simply by instinctively judging the timescale of echoes. Precise timing plays an important role in emotional responses as well, as any good comedian, film producer or musician will know.

One of the biggest challenges for scientists is that time perception occurs at many different levels – from milliseconds to years – and is very difficult to measure. A common approach is to ask individuals to estimate how long an event has lasted or how long ago something happened. This is a measure of remembered duration, which by definition is as much an assessment of memory as it is time perception. Alternatively, people can be asked in advance to make judgments, indicating when they think a particular time interval has passed. Rather than asking people to make retrospective and prospective judgments about the passing of time, we can assess time-related behaviour. For example, tapping and clapping tasks can be used to establish an individual's ability to create or maintain a regular pulse, and people can be asked to make perceptual judgments about whether two stimuli occur simultaneously or in sequence.

There is a whole host of experiments devoted to studying timekeeping in animals. Many of these take place in a laboratory. For example, one method involves presenting rats with food at regular intervals and then watching to see how well they are able to predict the next delivery of food. However, there is also more naturalistic research currently being undertaken: observations of hummingbirds, for example, reveal that they are very good at predicting the twenty minutes it takes for a flower to refill with nectar. Possibly the most intriguing time experiment was one carried out by the American neuroscientist David Eagleman in 2007. He was fascinated by the idea that everything seems to go into slow motion when we are in danger, so he and his team designed something called a perceptual chronometer, which presented electronic digits at different rates. First, he established the fastest rate at which digits could be registered by people in a relaxed state. Then, he strapped the device to their arm like a wristwatch, increased the rate of the digits by a small amount, and invited his brave participants to do a freefall jump from a platform that was fifteen storeys high. After they had landed safely in the net below, Eagleman asked them whether they had been able to read the faster-paced digits; no one could, which forced him to conclude that time does not actually slow down.

HOW FEELINGS AND THOUGHTS AFFECT TIME PERCEPTION

We will return to this concept and other strange time illusions, but let us first consider factors that influence our perception of time passing. One of the most familiar paradoxes is the feeling that time flies when we are having fun, and yet when we look back to the beginning of a very busy day, or over the first week of an exciting holiday, it can often feel like longer than normal. Think how many times you have heard someone say, *It's been a long day!* On

Above: How do you picture a month in your mind? Personality, language, experience and culture all influence how we see time.

the other hand, when we are doing something very boring or monotonous, time really seems to drag, and yet at the end of the week those dull days can actually seem as though they were the shortest. A popular explanation for this phenomenon relates to attention: during really busy and exciting times, most of our brain resources are taken up, leaving little opportunity to process the passing of time. In contrast, when we look back over this period we have a sense of many events having been processed, which gives us the illusion that *more* time must have passed.

We also seem to make different time judgments for things we see, compared to those we hear: on the whole, sounds seem to last longer than pictures, even if they are presented for the same length of time. According to Sylvie Droit-Volet of Clermont University in France, our emotional state also impacts on time judgments. She found that people tend to feel as though they have been looking at emotional faces for longer than neutral ones, but she believes this is because we take on those emotional states ourselves. To test this hypothesis, she repeated the experiment but asked people to hold a pencil horizontally in their mouth, which forces them to smile. Under these conditions, in which people were "smiling" as they looked at *all* of the faces, there was no longer a difference between the duration of analyzing emotional and neutral faces.

HOW DOES THE BRAIN TELL THE TIME?

Differences in cultural background and personality have a bearing on the perception of time, as well. When you think about a year, how do you picture it? Is it a circle or wheel, or one big rectangle? Or do you see it as twelve separate pages, like a calendar? What about a week? Is it an undulating line with weekends at the top and Friday at the bottom, or perhaps a never-ending ladder? When you ask someone to put a deadline *forward a week*, do you mean a week later or a week earlier? There are big individual differences in people's answers to these questions and personality plays a part in this. Research shows that mental representations of time can influence the way we talk about it. Many of the words we use to describe time are borrowed from those we use to describe space – for example, we talk about events occurring before and after each other, closer or further away from now, and of being a particular length. In fact, time and space are so intricately linked that Francesca Frassinetti, of Bologna University in Italy, was able to demonstrate that wearing prism glasses that visually shift everything to the right makes us *over*-estimate time durations; conversely, left-shifting glasses make people *under*-estimate time durations.

PHYSIOLOGICAL FACTORS

Our physiological state can also affect our perception of time, as anyone who has experienced a high fever may know. An early experiment in this area was carried out by the American physiologist Hudson Hoagland, who became intrigued by aspects of time perception while caring for his wife during a bout of influenza. Each time he went to her bedside, she complained that he had been away too long; he wondered if she was simply judging time differently to him, due to her feverish state, and decided to test this theory. He invited her to estimate the length of a minute while he timed her on his watch and found that she consistently under-estimated by up to 25 seconds. Time really did seem to pass more quickly for her. Of course, this may in part be explained by the fact that being ill is boring and monotonous, but many experiments have since supported a direct link between body temperature and time perception: as we get hotter, we perceive time to go faster, and vice versa when we are cold.

Below: David Eagleman carried out experiments to see whether time really slowed down for people jumping from a great height.

rightening free-falling experiment (*see* page 100)? Despite he fact that time did not objectively go into slow motion for his participants, they were still very prone to over-estimating how long the jump had taken. Eagleman and his colleagues uggest this relates to attention and arousal – the sudden boost of adrenaline heightens all our senses and puts the brain on super alert, meaning that it processes many things very quickly. Just as with the feeling that we have had a long day when we have been busy, this gives people a sense that he period of danger was much longer than in reality.

British television satirist Chris Morris amused his audience by successfully persuading a number of media figures that a new drug, "cake", had a specific effect on the time perception area of the brain, which he jokingly referred to as Shatner's Bassoon (this being an oblique reference to the actor William Shatner, who played Captain Kirk in *Star Trek*, and the faintly comical sounding name of a musical wind instrument). Although this was a good example of how easily people can be seduced by jargon and neuroscience, the drug and the brain anatomy were both of course complete fiction Nevertheless, there is good evidence from studies featuring both humans and animals that recreational drugs do hav an impact on time perception. Smoking cannabis seem to reliably stretch perception of time, which some argue can enhance the experience of certain time-based activities especially music. Conversely, cocaine, methamphetamine and alcohol tend to increase the perceptions of the speed o time, while hallucinogens such as psilocybin have a more complex effect, shortening the reproduction of time interval but slowing down the pace of a freely chosen rhythm. All o these effects suggest that simple biological changes can tampe with our internal clock. Certain neurological and menta health conditions also cause disruption to time perception particularly Parkinson's disease and schizophrenia, but also depression and Attention Deficit Disorder.

BRAIN CLOCKS

So, what exactly is this "internal clock" and where is i in the brain? Do we even really have one? At the age o just 23, the French speleologist Michel Siffre carried out a very challenging investigation of his internal body clock He spent two months in total isolation in an underground cave in the French alps with no light, no watch or clock and therefore no idea of whether it was night or day. He attempted to follow his own natural rhythm – sleeping waking and eating, as and when it felt natural. He would make a phone call to his collaborators when he woke and again when he went to bed, and they monitored hi habits. He was already two hours out of sync by the second morning and when he was told 62 days later that the experiment had finished, he believed he still had 25 day to go! Until this point, no one knew the natural length o the human sleep/wake cycle, but analysis of Siffre's habit revealed a very reliable pattern that was 24½ hours long Subsequent research indicates that this now appears to b typical of other human beings.

Michel Siffre's subterranean expedition provided a important foundation for our understanding of human

Below: Psilocybin, found in "magic mushrooms", an alter the sense of time passing.

arousal, hormones, immune function, body temperature and digestive activity that is governed by a small area in the hypothalamus named the suprachiasmatic nucleus (or SCN). Neurons in this tiny region of the brain fire nerve impulses in a regular pattern of just over 24 hours – generally fastest at midday and slowest at night. There is a direct input to the SCN from the eyes, which feeds in information about whether it is night or day and, as we know from Siffre's research, corrects the natural internal clock so that it is in sync with our environment. When the daylight signals lead to drastic alterations – for example, when people work nightshifts or travel by air to different time zones – it causes significant disruption of these natural physiological rhythms, which explains why jet lag can leave people feeling quite unwell.

Right: Some of us naturally function better in the morning – the "larks". Others, the "owls", are better in the evening and at night.

Below: Our brains follow a natural circadian rhythm, which is influenced by light and dark.

Once we explore further and get beyond the SCN, our understanding of the brain's timing mechanisms becomes less detailed. This neat mechanism may explain how we regulate long cyclical periods of time, but what about those millisecond judgments that are so crucial to communication, accurate perception and coordinated behaviours? One region of the brain that seems to be particularly important is the cerebellum – the large, cauliflower-shaped structure that hangs off the back of the organ. We know that this plays a crucial role in coordinating movements and detecting musical rhythm, so it is perhaps not surprising to find that it also seems to be the place where we process very short time intervals. People with damage to the cerebellum have difficulties with timekeeping and other timing tasks, and the importance of its role has been backed up with neuroimaging studies in healthy people.

Two other important brain areas are the basal ganglia and the prefrontal cortex. In 2001, the American psychologist and neuroscientist Warren Meck found that the basal ganglia, a loop of structures towards the middle of the brain, are responsible for managing slightly longer timescales, probably those that are more than one or two seconds but less than a few minutes long. Again, this area is important in the control of movement and it is also damaged in people with Parkinson's disease. Meck

Below: The cerebellum plays an important role in the split-second timing necessary for coordinating movement.

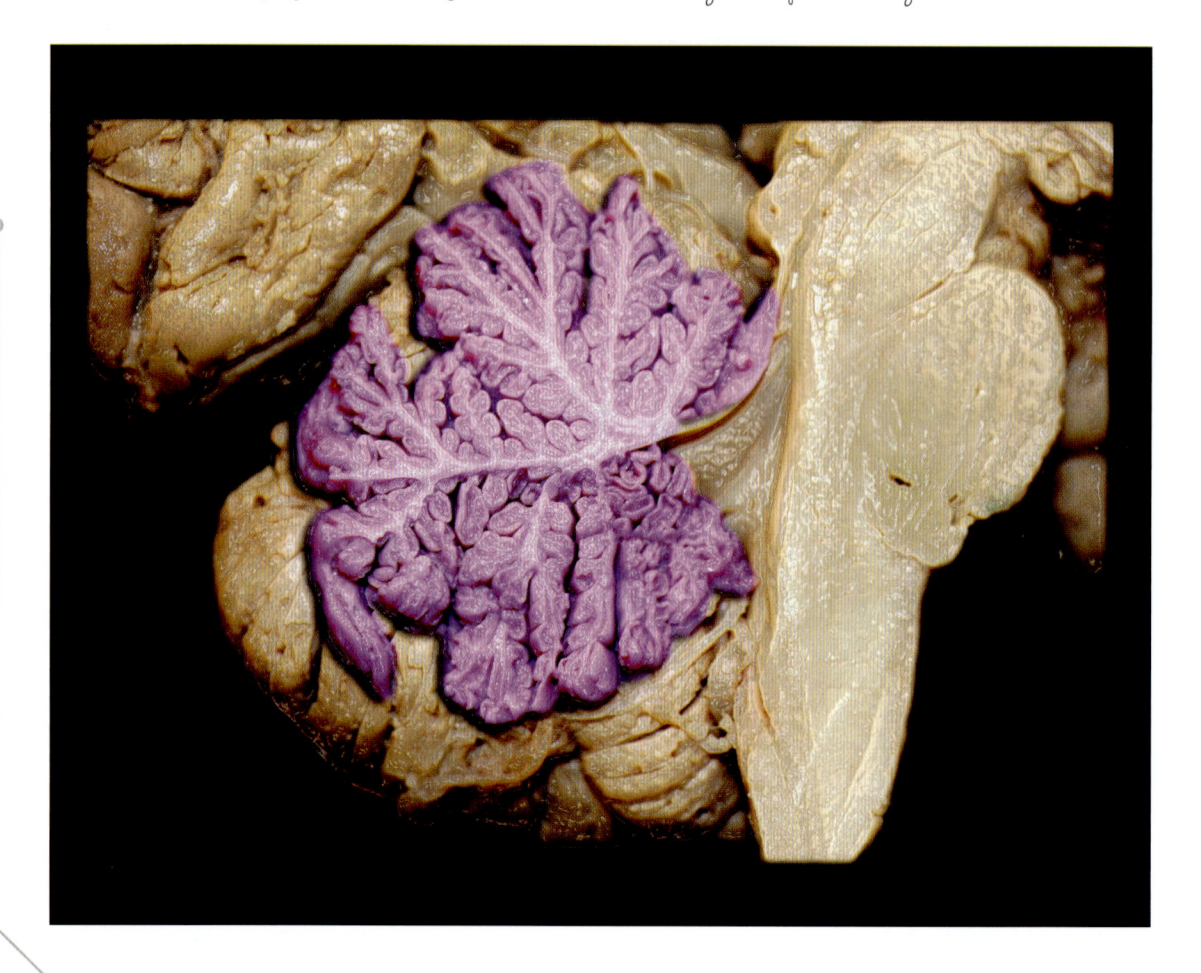

HOW DOES THE BRAIN TELL THE TIME? 105

Fig. 8.1: Areas of the brain related to time perception

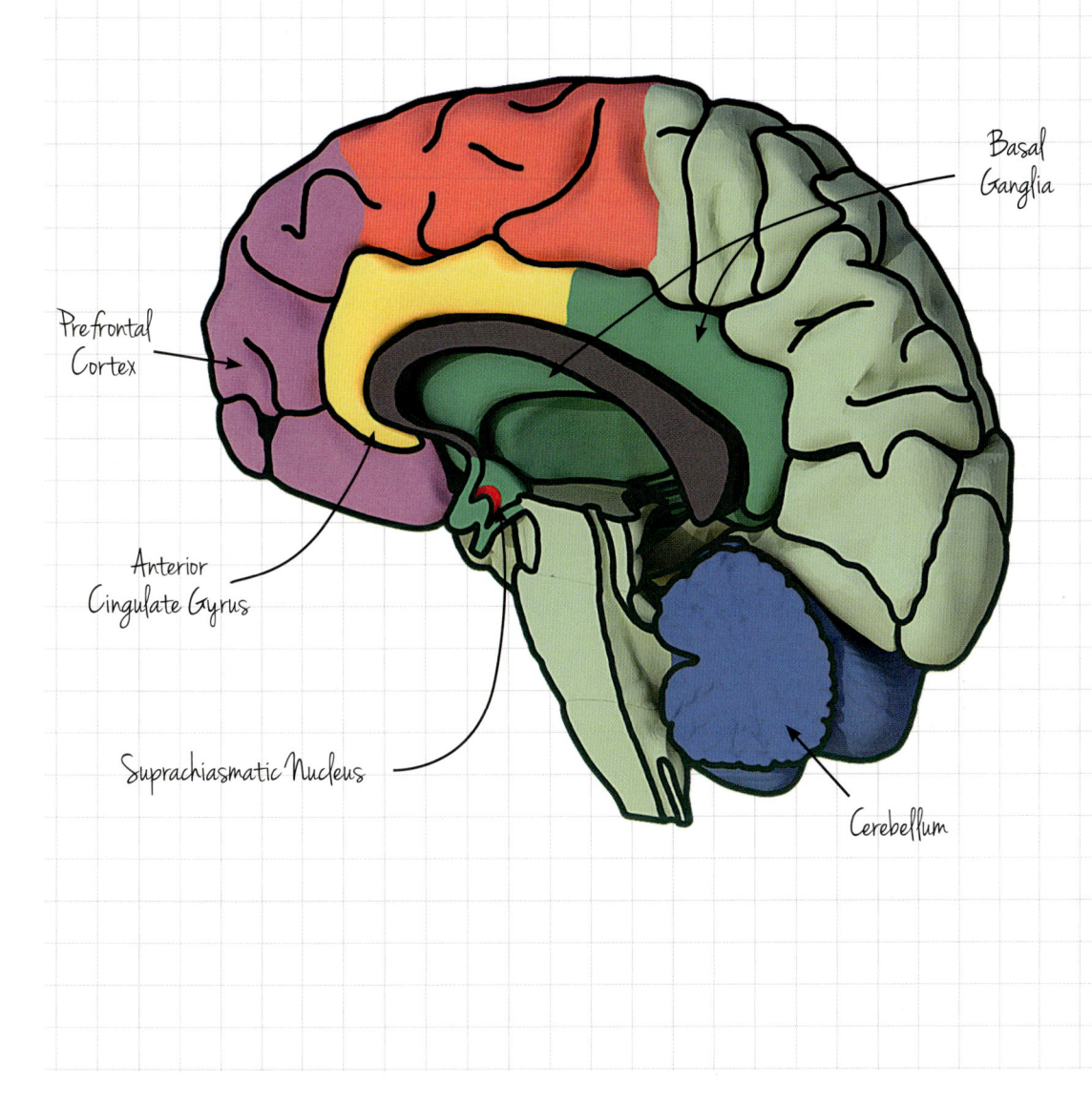

- Basal Ganglia
- Prefrontal Cortex
- Anterior Cingulate Gyrus
- Suprachiasmatic Nucleus
- Cerebellum

has proposed that the basal ganglia detect the natural oscillations from different body organs and use this data to form a clock, which assesses time intervals by adding up the number of pulsations.

The smaller scale timing abilities managed by the cerebellum and basal ganglia are relevant to many day-to-day activities, but arguably distinct from our broader conscious awareness of time passing – the feeling of *now* compared to *then* – and longer-term predictions of the future. This is more likely to be managed by the prefrontal cortex (PFC), at the very front of the brain. The PFC is also the seat of our working memory, which binds together, organizes and sequences our many different thoughts and sensory inputs. It is probably this capacity that gives us the sense of being in the moment and dictates how that relates to what has come before or is likely to come next. This region also controls attention and is affected by recreational drugs, both of which we have seen can interfere with time perception.

TIME AND MEMORY

So, we know from brain-imaging studies and investigations with patients that the SCN, cerebellum, basal ganglia and PFC are all important structures for perceiving and processing time. However, it is likely that others are important, in addition – for example, memory is so inextricably bound up with time that the hippocampus must also be relevant. One clue that we use to establish how long ago something happened is the strength of the memory in question – the more vivid it is, the more recently we assume it has occurred. This can be misleading, of course, because a particularly powerful memory will have a stronger trace and therefore often feels like it was more recent than it actually was. This explains why you may sometimes be surprised at how long ago an important event took place.

How all of this actually works remains a real mystery. We know that all of the brain structures listed above are

TOP TIPS FOR TIME MANAGEMENT

"*How did it get late so soon?*" wrote the American children's author Dr Seuss – a common complaint in this day and age. In fact, time management is a concern of such widespread magnitude that entering these words as the search term on an online bookstore brings up nearly 58,000 results! If you don't want to read a whole book on it (or don't have time), here are a few practical scientific tips on how to organize time better:

1. Develop routines, use calendars and make lists – the more habitual and organized we are, the less we waste precious brain resources on making trivial decisions;
2. Know your limits – attention is a finite resource, so don't try to spread it too thinly;
3. Before you go to bed, visualize the five most important things you want to do the next day – imagining future events activates the relevant brain areas and makes you less likely to procrastinate;
4. Get in sync with your body clock – we are all better at doing different things at different times of the day; for example, you may be more alert in the mornings and more creative in the evenings, so try to identify and work with your natural pattern;
5. Alter your perception of time – Claudia Hammond, author of *Time Warped: Unlocking the Mysteries of Time Perception*, suggests that few time management techniques will help everyone and that a better approach may be to change how we use and remember our time. Time diaries show that busy people have often had more free time available to them than their memories would have them believe. And when estimating how much time a task will take, Hammond tells us to break it down into individual chunks or, better still, ask someone else to estimate for us. They are almost always going to be more objective and realistic in their assessment.

involved, but what do they actually do? A popular long-term theory has been the idea of a neural clock that sends out regular pulses, which are counted up by an accumulator and then stored or compared with previous inputs. Although this principle works well with a lot of the empirical findings – for example, the clock may speed up or slow down with changes in dopamine levels or temperature, and the accumulator may work differently according to varying levels of attention – it does not fit neatly with the anatomical knowledge that we have. Professor Warren Meck's recent experiments suggest that the basal ganglia and striatum may play this role to some extent, but how do we account for the apparent neural distinction between different timescales?

Most theorists now accept that all aspects of timing rely to some extent on the natural oscillations found in the brain and probably require a combined network of these key brain regions. However, there is still some way to go before we fully understand the mechanisms involved. And despite our advanced timekeeping skills, humans clearly have an inherent need to measure time objectively – an enterprise that goes back to at least 2000 BC and which has led to the development over the centuries of extraordinarily accurate clocks. History has shown us that removing or deliberately altering all external time cues for an individual – clocks, mealtimes, light and dark – can have an exceptionally detrimental effect on their mental health and is even considered to be a form of torture. Much of the confusion that occurs in dementia patients and amnesiacs comes from not being able to correctly sequence their few remaining memories – did it happen yesterday or twenty years ago? The only certain conclusion to draw is that whatever it is, and however we measure it, time is vitally important to us all.

Below: We rely on feelings of vividness to judge how old a memory is – this ability often fails in people with dementia.

9. MUSIC IN THE BRAIN

CAN YOU IMAGINE LIFE WITHOUT MUSIC? FOR MOST PEOPLE THIS IS A DREADFUL THOUGHT. MUSIC IS ALL AROUND US - IN SHOPS, RESTAURANTS, HOMES, CARS, CONCERT HALLS, CHURCHES AND ON FOOTBALL TERRACES.

In recent times, music is even more accessible: people are able to plug a pair of headphones into a portable player and carry their own private musical worlds with them. In fact, our own research recently showed that on average, people hear music, either for real or in their head, a staggering 75 per cent of the time! And music is extraordinarily powerful. It can cheer us up, make us cry, lull us to sleep, urge us to dance, lure us into an embrace or influence what we buy. Research has even found that the right music can improve our immune response and affect how quickly wounds heal.

According to the Chinese philosopher and teacher Confucius (551–479 BC), *Music produces a kind of pleasure which human nature cannot do without.* Historical evidence supports this view: ancient cave paintings depicting use of old-fashioned flutes and drums suggest that music has always been a central part of human life. This must in part be because of its unique ability to influence and moderate our moods. A recent piece of research by Patrik Juslin and Petri Laukka, researchers in musical psychology at Uppsala University in Sweden, found that people have a very broad vocabulary when it comes to describing their

Below: Music has the power to coax us into buying romantic gifts, such as flowers, as well as to lull us to sleep.

emotional responses to music. Commonly used words range from simple concepts such a, *happy*, *sad* and *calm*, through to more complex emotions including *honoured*, *jealous*, *curious* and *empathic*. So why is music so powerful and what is going on in the brain when we listen to our favourite tunes?

HOW DOES MUSIC AFFECT YOUR BODY?

Although the modern concept of music therapy did not really emerge until after the Second World War, music was already being used in hospitals for its relaxing and entertaining qualities, to boost patient morale and aid convalescence. In fact, evidence shows that music has been used therapeutically for as long as any kind of records have existed, and it is still used in the healing rituals of many tribal cultures today. A recent meta-analysis from Jayne Standley, an American professor of music therapy, showed that in young babies who had just undergone surgery, music significantly reduced their pain, respiration rate, pulse, blood pressure and the need for analgesia. Powerful stuff! Other research has shown that listening to music can reduce pain perception in those suffering from cancer and alleviate allergic reactions in some patients with skin conditions.

It is very impressive to think that music can have such a measureable impact on the human body – but we do not have to be ill in order to experience physical effects. Most of us will be familiar with the shiver down the spine that happens when we hear music that really moves us. This feeling has been dubbed the *chill response* and the majority of people are able to easily identify the moment when it happens, making it a relatively straightforward phenomenon to investigate. Music that gives us this sensation has a very real effect on the body: our heart rate increases, we breathe more quickly, our body temperature changes.

The chill response also boosts levels of the "cuddle hormone" oxytocin, which is linked with feeling connected and loving toward other people. This may explain why going to a musical concert or singing together can be such a socially bonding experience. Interestingly, it may also explain why music is a powerful marker of cultural identity,

Above: Singing together, for example at a sporting event, boosts levels of oxytocin in the brain, which may explain why it is such a strong form of social bonding.

sometimes to the point of being divisive. The famous distinction between British "mods" and "rockers" in the mid-late 1960s and into the 1970s – which led to some highly aggressive conflicts and sparked fears of the breakdown of youth culture – was essentially defined by whether people listened to Eddie Cochran and Gene Vincent or the Who and the Small Faces. Although described as a hormone of empathy and bonding, oxytocin has also been shown to increase aggressive feelings and hostility toward those who are not part of the group. So, while music might make us feel more connected to our sub-culture, it may also cause us to feel more distant from those who are not part of it. This may be especially true during adolescence, when identity formation is at its peak and when we are generally at our least open-minded when it comes to which music we choose to listen to.

Oxytocin is not the only neurochemical that is affected by emotional music. The Estonian-American psychologist Jaak Panksepp and others have also found that music can cause an increase in endorphins, the body's own natural form of heroin or morphine. These chemicals help us to deal with pain and give us feelings of pleasure and contentment by stimulating the reward pathways in the brain. Dopamine levels are also boosted by some music, adding to the activity of these "feel-good" systems, which may explain why many tunes are so compelling. In fact, there is now an abundance of evidence to prove that music activates not just these areas of the brain but also many others. But which brain structures exactly play the most important roles in understanding and producing music?

MUSIC IN THE BRAIN

In 1953, at the age of 51, the Soviet composer Vissarion Shebalin suffered a stroke. This was followed by a second more serious stroke six years later, which left him partially paralyzed on his right side and with severe disturbances to his speech and writing. Despite this, he continued to write music, and his fifth symphony, composed just three years after his second stroke, was described by his friend Shostakovitch as *a brilliantly creative work, filled with highest emotions, optimistic and full of life*. A post-mortem revealed massive damage to the left parietal and temporal lobes of Shebalin's brain.

This case has sometimes been used, along with other evidence, to suggest that music is a right-brain function in contrast to language, which is generally controlled by the left side of the brain. However, a few years after Shebalin's post-mortem, two researchers named Wertheim and Botez reported a very detailed study of a professional violin player who also had significant damage to the left side of his brain. They found that although some of his musical skills were intact, other musical abilities were quite seriously impaired. For example, the violin player no longer had perfect pitch, was unable to recognize familiar pieces of music and struggled to read sequences of notes.

What we now know is that listening to, creating and performing music are very complex skills that involve many different parts of the brain – and we certainly cannot say that music lives in the right side of the brain. A particularly compelling piece of evidence comes from a study carried out by Lawrence Parsons and colleagues in Sheffield, England. They used PET scans to examine brain activity in a group of pianists while they were performing a concerto by Bach. The neural activations were so complex and widespread that it was almost easier to list the areas of the brain that were **not** active than to list all those that were! This is not surprising, given all of the cognitive and emotional processes that are likely to be involved in understanding, responding to and producing a piece of music.

EEG AND SPECIFIC BRAIN REGIONS

There are also some intriguing studies that have analyzed electrical activity of the brain using an electroencephalogram (EEG). For example, researchers at the Max Planck Institute had the fascinating idea of measuring brain activity in eight pairs of guitarists while they were duetting together. Almost as soon as the lead guitarist had signalled when to start, the brain activity of each pair became synchronized, demonstrating that the same sets of neurons were firing and in the same rhythm in all the guitarists' brains. Other research has suggested that music that has a particularly strong and repetitive beat – for example, traditional African drumming, 1980s style Acid House or modern electronic dance music (EDM) – might also lead to very regular, synchronized brain waves. This finding is probably reinforced by the fact that these kinds of music usually provoke dancing or moving in time to the beat.

There are several regions of the brain that seem to be particularly important when it comes to emotional responses to music. Predictably, these include the limbic system and amygdala (*see* Chapter 7), but recent work by the Canadian researcher Dr Robert Zatorre has also found very specific activation of the reward pathway that directly correlates with experiencing the chill response. People undergo a short period of anticipation when they know that they are about to experience an intense moment, and at this point the reward response starts to kindle in the caudate nucleus. Then, at the moment of climax, there is a wave of activation that spreads up to the nucleus accumbens, the brain's pleasure centre. It seems that rock 'n' roll really does act on the same systems as sex and drugs.

Another brain region that sees a lot of activity stimulated by music is the *auditory cortex*, which receives and interprets sound. There are particularly strong responses in the areas that are sensitive to the emotive elements of communication, that is, laughter, screams and prosody of speech. There is also a small area, named the *parahippocampal gyrus*, which seems to process dissonance – sounds that seem to clash and sound unpleasant together. Finally, there is increased activity in the *cerebellum*, which seems to drive the overwhelming

MUSIC IN THE BRAIN 111

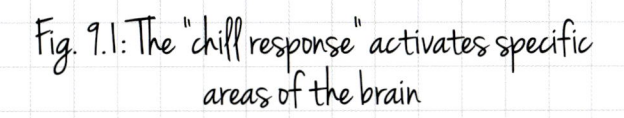

Fig. 9.1: The "chill response" activates specific areas of the brain

1. Anticipation of musical climax activates caudate nucleus

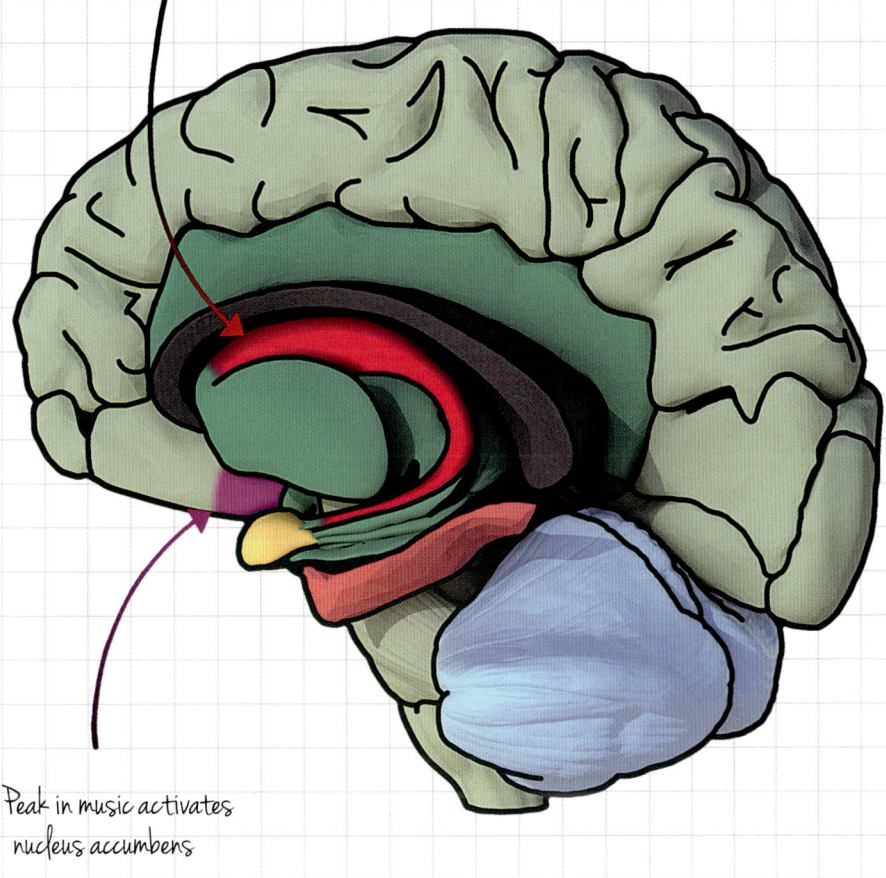

2. Peak in music activates nucleus accumbens

Fig. 9.2: Emotional response of the brain to music

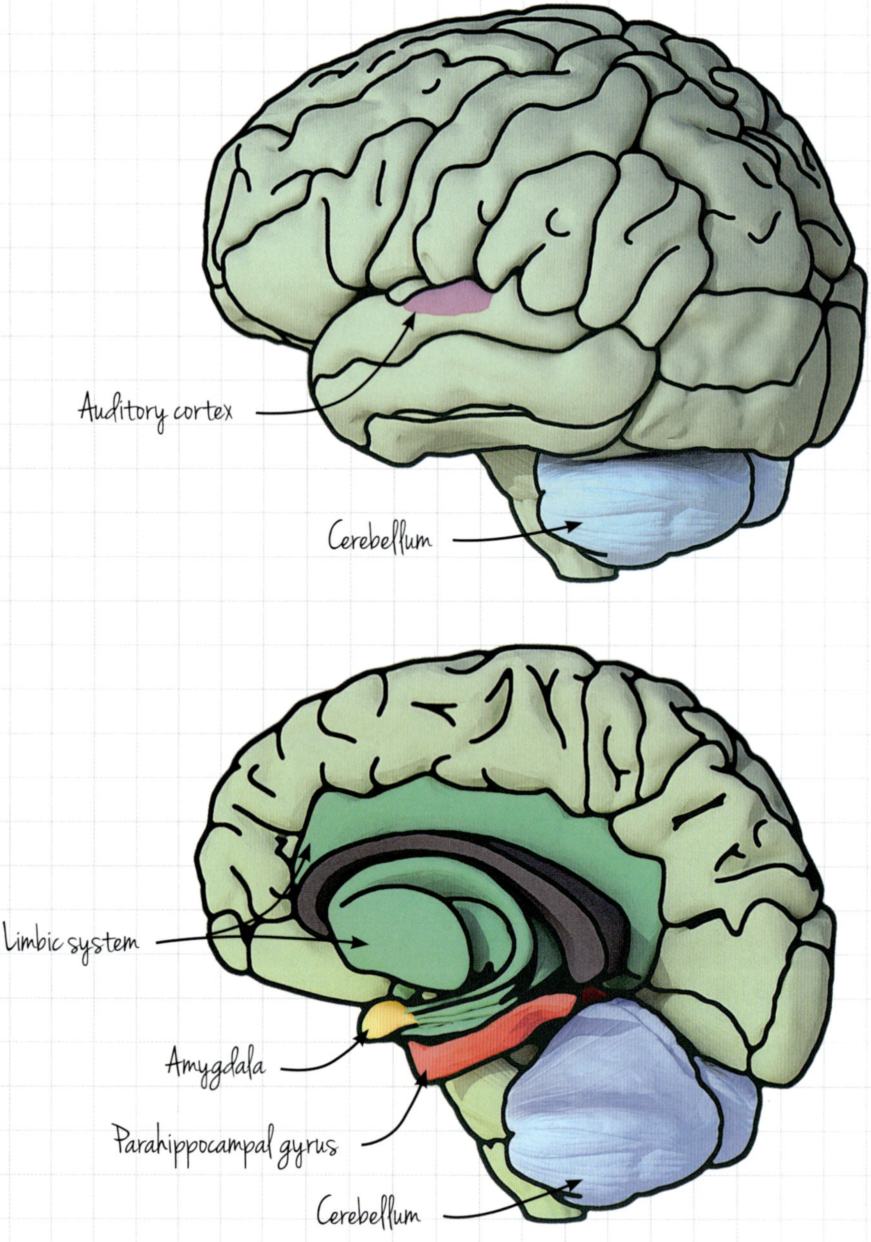

urge people have to tap their feet, sway in time or get up and dance when they hear a good song.

THE EFFECT OF MUSIC ON ANIMALS

It seems that the power of music is not limited to influencing human beings – dogs, cows, rats and even chickens, have been shown to respond to music. Adrian North, a professor of psychology at the University of Leicester in the United Kingdom, carried out a study in which he examined milk production in cows that were listening to different types of music, played for twelve hours a day. North and his colleagues found that slow music, such as the American band REM's 'Everybody Hurts' or Beethoven's Pastoral Symphony, increased the milk yield by 0.73 litres (1.5 pints) per cow compared to when they were listening to faster songs such as 'Back in the USSR' by the Beatles or (yes, you've guessed it) the Wonder Stuff's 'Size of a Cow'…. They hypothesized that the slower music helped to alleviate stress and relax the cows.

Another study looked at the behaviour of 50 dogs in a rescue shelter while different types of music were piped through the speakers. They compared classical music, heavy metal music, pop music and human conversation. Not surprisingly the dogs were more calm, rested and relaxed when the classical music was being played, whereas heavy metal seemed to provoke a noisy chorus of barking. It has even been shown that music can affect health measures in animals. Harp music helped to slow the respiration rate of dogs that had just undergone surgery, and conversely, loud rock music slowed the rate of wound healing in a group of rats.

What is particularly intriguing is that some animals show signs of that shiver-down-the-spine experience mentioned earlier. In a landmark experiment, Jaak Panksepp (*see* page 109) found that even baby chickens experience the chill response by fluffing up their feathers, shaking their heads and releasing oxytocin. What was the music that provoked this reaction most effectively? Pink Floyd's 1983 album *The Final Cut*. Panksepp uses this study and others to suggest that the evolutionary basis of music is ancient and relates to primal communication of important needs, such as the sounds made by crying offspring.

Below: Baby chickens fluff their feathers, shake their heads and release oxytocin in response to certain types of music.

WHY IS MUSIC SO POWERFUL?

The question of why we have evolved to be so emotionally responsive to music and seem to need it so much is intriguing. What is it about a sequence of organized sounds that can create such an overwhelming and compelling aesthetic experience? Panksepp's evidence and theories suggest that music taps into a very fundamental system related to communication and bonding and intuitively this makes a lot of sense. All animals, humans included, use the elements of music – pitch, timbre and rhythm – to add emotional content to any communication. We might use all manner of expletives to indicate to a dog that it has misbehaved, but it will be our tone of voice that really conveys the message. Likewise, long before a baby can speak, we use musical sounds to soothe, excite or reprimand.

Humans are able to hear long before they are born – from about twenty weeks after conception – so unborn babies are exposed to everything from their mother's laughter and crying to the theme tune of her favourite television show. In fact, researchers have demonstrated that newborn infants are able to show recognition of music they have heard in the womb. Other studies have demonstrated that babies are able to identify different rhythms in the first few days of life and are even able to learn new melodies. Therefore, we certainly have a strong sensitivity to musical sounds from the moment we are born.

Jaak Panksepp argues that the use of pitch, timbre and rhythm in our communications is central to our social behaviours and attachments. This includes sounds of laughter, crying, screaming, whining, soothing, and even the sounds that are produced when animals mate or humans make love. These are all examples of pre-linguistic communication – that is, they do not require words and can be used effectively by all species at any age. But Panksepp also talks about paralinguistic elements of communication – the way that we change the pitch, loudness and tone of our voices to convey particular meaning. Most of us have occasionally got into difficulties because of misunderstanding a piece of written text, such as an email or an SMS message, because it is easy to lose the precise meaning of words when they do not have the nuances of the audible human voice behind them. So, in a nutshell it is pitch, timbre, melody and rhythm that allow us to convey emotion in speech, and this is probably why we are so innately receptive to music.

Left: Music is a powerful marker of cultural identity.

Right: Studies suggest that newborn infants are able to identify rhythms in the first few days of life.

PITCH, TIMBRE, RHYTHM AND MELODY

The term *pitch* refers to how high or low a sound is – a scream has a high pitch and a rumble of thunder a low pitch. The word *timbre* is used to describe the tonal characteristic of an instrument or voice – a flute has a different timbre to a violin and your voice will sound different to mine. *Rhythm* describes the underlying pulse, timings and stresses of different sounds in relation to each other – for example, *hippopotamus* and *hot cup of coffee* both contain five sounds, but the rhythm of each figure is different. A *melody* is a sequence of notes – in music this might be called a tune, but in speech it is describesd as *prosody*, referring to the way that our voice goes up and down when we speak.

HOW DOES MUSIC CREATE EMOTION?

So music makes us dance, cry and shiver with joy, because it taps into the brain systems that allow us to connect with other people. However, the million-dollar question for every composer and record producer is why some tunes seem to be so much more emotive and popular than others. Why is it that some melodies get stuck in our heads for hours, or even days on end, and what is it that characterizes those musical moments that cause that magical *chill response*? Most intriguingly, why is it that the same piece of music can leave some people cold and reduce others to tears?

An obvious suggestion is that some songs remind us of particularly poignant times – a wedding day, a first dance, a funeral, or maybe more general memories from a difficult or exciting period in our life, such as starting at university. This notion of consciously associating pieces of music with our past was first put forward by the scientist John Booth Davies in 1978 and has been affectionately called the *Darling, they're playing our tune* theory. The huge popularity of *Desert Island Discs*, BBC Radio 4's long-running programme, is testament to the public's fascination with musical memories. However, memory can also contribute to our feelings for music that we have never even heard before. From an early age, we become conditioned to associate particular emotions with particular musical structures. For example, in Western culture we learn that minor harmonies usually signal sadness and major harmonies, happiness. We can also "catch" particular emotions from music, if it mimics social communication sounds such as laughter or crying.

A VARIETY OF FACTORS

Memory alone is not enough to account for the richness or diversity of the emotional reactions that people have to music, so what else might come into play? In 2012, Britain's new pop sweetheart Adele swept the board with

Below: Music is a powerful communication tool, capable of provoking joy and sadness.

Above: Combining music with dance can have a particularly powerful emotional effect.

six Grammy awards for her achievements in the music industry. One of her big hits – 'Someone Like You' – has been described as the perfect tear-jerker and was voted the third most popular Number 1 single of the last 60 years, coming behind only Queen's 'Bohemian Rhapsody' and Michael Jackson's 'Billie Jean'. Has Adele therefore hit upon a winning formula?

Twenty years earlier, British music psychologist John Sloboda carried out an in-depth study in which he asked people to locate the precise moments in different pieces of music that gave them an intense emotional reaction. He, and others after him, has identified particular musical structures and events that are good at provoking emotion. For example, Sloboda found that music which speeds up or has a syncopated beat – the kind of uneven rhythm that occurs in Latin music or a lot of modern pop – seems to be particularly arousing and leads to a racing heart. He also discovered that sudden changes in harmony provoke the chill response and that people feel tearful when they hear a build-up of harmonic tensions that then resolve.

The real key, according to Sloboda, is the way that composers and performers create and fulfil expectations in the listener. This is exactly the same principle by which a good story told by a funny comedian works – as the story or joke unfolds, we begin to anticipate where it will go next and how it will end. Some resolutions will give us a lump in the throat or make us feel tearful, while others may give us a sense of excitement or surprise. Even something as simple as a musical change of key can have a very powerful and emotive effect, as pop music producers have long known. Adele uses the expectancy device very effectively in her song 'Someone Like You'; the verse is gentle, soft and repetitive, and then when the chorus comes in she suddenly jumps up the octave to a much bigger, stronger sound. There is also a moving shift in harmony.

Above: Much like music, slapstick humour relies on the element of surprise to provoke an emotional reaction.

As with humour, this concept of building and then meeting expectations can happen at a very simple level – a sudden change in speed or volume, for example – and will provoke a similar emotional reaction in people of all ages and cultures. Conversely, it may be more complex and might require a more developed, culture-specific understanding of music. In such a case, the responses will be richer and more intense, but also more dependent on individual taste and experience. An analogy in comedy is a comparison between slapstick artists such as Laurel and Hardy or the typical circus clown, and comedians like Ricky Gervais or Woody Allen, who will only evoke humour in those who understand and relate to the stories they tell. This is why, despite Adele's success, her music does not appeal to everyone. Essentially, if she is not speaking our musical "language", then there is a limit to how moved we will be.

Music has parallels with language in that it is a way in which we can communicate and bond with others; it shares many of the same brain structures and it takes on very different forms across different cultures. We cannot use it to express concrete ideas such as *the weather looks nice today*, but there is no doubt that it communicates *something* and that it does so very powerfully. Stephen Pinker, the Canadian-American popular science author and cognitive scientist, argues that music is nothing more than *auditory cheesecake*. By this he means that it hijacks the fundamental elements of our communication system and puts them together in a form that gives us pleasure. Jaak Panksepp disagrees with this theory and proposes that music is an intrinsic and important part of our evolution – it provides us with a unique opportunity to bond and connect with others. Either way, music is strongly woven into the fabric of our human lives and will probably always be so.

CASE STUDY: CLIVE WEARING

In March 1985, Clive Wearing (*see* page 65) was rushed into the Accident and Emergency department of St Mary's hospital in Paddington, London, with a high fever, severe headache and acute confusion. He was diagnosed with encephalitis, a viral inflammation of the brain. Doctors were able to halt the spread of the virus and save Clive's life, but not before it had destroyed large areas of his brain, leaving him with one of the most severe forms of amnesia ever documented. At the time of his illness Clive was an intelligent, passionate and highly accomplished musician who directed a professional choir and worked for BBC Radio 3. The damage to Clive's brain wiped almost all of his memories and left him unable to make any new ones. Despite the fact that Clive lives entirely in the moment and needs full-time care, his musical skills have remained largely untouched. He can still play the piano brilliantly and hum along to music he knew from his past. Most importantly, his passion and love of music remain a core part of his life. His case is just one of many that illustrate the resilient and fundamental nature of music.

10. THE YOUNG AND DEVELOPING BRAIN

THE HUMAN BRAIN IS A REMARKABLE ORGAN - IT STARTS LIFE AS A SMALL BALL OF CELLS AND WITHIN MONTHS HAS BECOME THE VERY SEAT OF OUR CONSCIOUSNESS. HOW?

As any new grandparent, aunt or uncle very quickly finds out, buying a gift for a baby can be a somewhat daunting task in today's western world. Old-fashioned rattles and teddy bears have been replaced by rows of exciting-looking "smart toys", each of which offers its own unique promise for boosting brain development. There are baby mobiles to hang above the cot, featuring complex black and white geometric patterns designed to stimulate the visual cortex; specially recorded versions of Mozart's 'Eine Kleine Nachtmusik', to promote early language development; or noisy electronic gadgets that encourage advanced recognition of numbers and letters. But do any of these make any difference? Can you really make a baby more intelligent and better behaved by surrounding it with scientifically developed educational toys? Or is it enough simply to feed, love and care for a child?

Above: In the Baby Lab in Geneva, a researcher tests babies to determine how language learning works.

GROWING THE BRAIN

We saw in Chapter 5 that memories are formed by changes to the synapses and the development of cell networks. The concept that our brains are dynamic organs which are physically shaped by our experiences is known as *plasticity*. As far back as 1890, William James, an American psychologist and philosopher, suggested that the brain was not completely fixed and was capable of reorganizing itself to some extent. It was a few decades later that physiological studies began to back this assertion up properly, and in 1948 the term *plasticity* was officially introduced by the Polish neuroscientist, Jerzy Konorski. We now know that although genes play an important part, our brain structure is heavily influenced by a whole range of environmental factors – nutrition, intellectual and sensory stimulation, toxins, love and attachment, infections, type of birth and stress, to name but a few – and this has important implications for social, educational and medical policies.

There are four processes involved in building the brain: these are *neurogenesis*, *synaptogenesis*, *myelination*, and *pruning*. As the name suggests, neurogenesis refers to the birth of new neurons and glial cells. Until very recently, it was believed that in humans neurogenesis happened only during gestation, and that by the time babies were born they had their full quota of neurons. However, in 1998, a landmark piece of research showed that the adult human brain was capable of producing new neurons in the hippocampus, the region of the brain important for memory. Subsequent studies have found that we are also

THE YOUNG AND DEVELOPING BRAIN

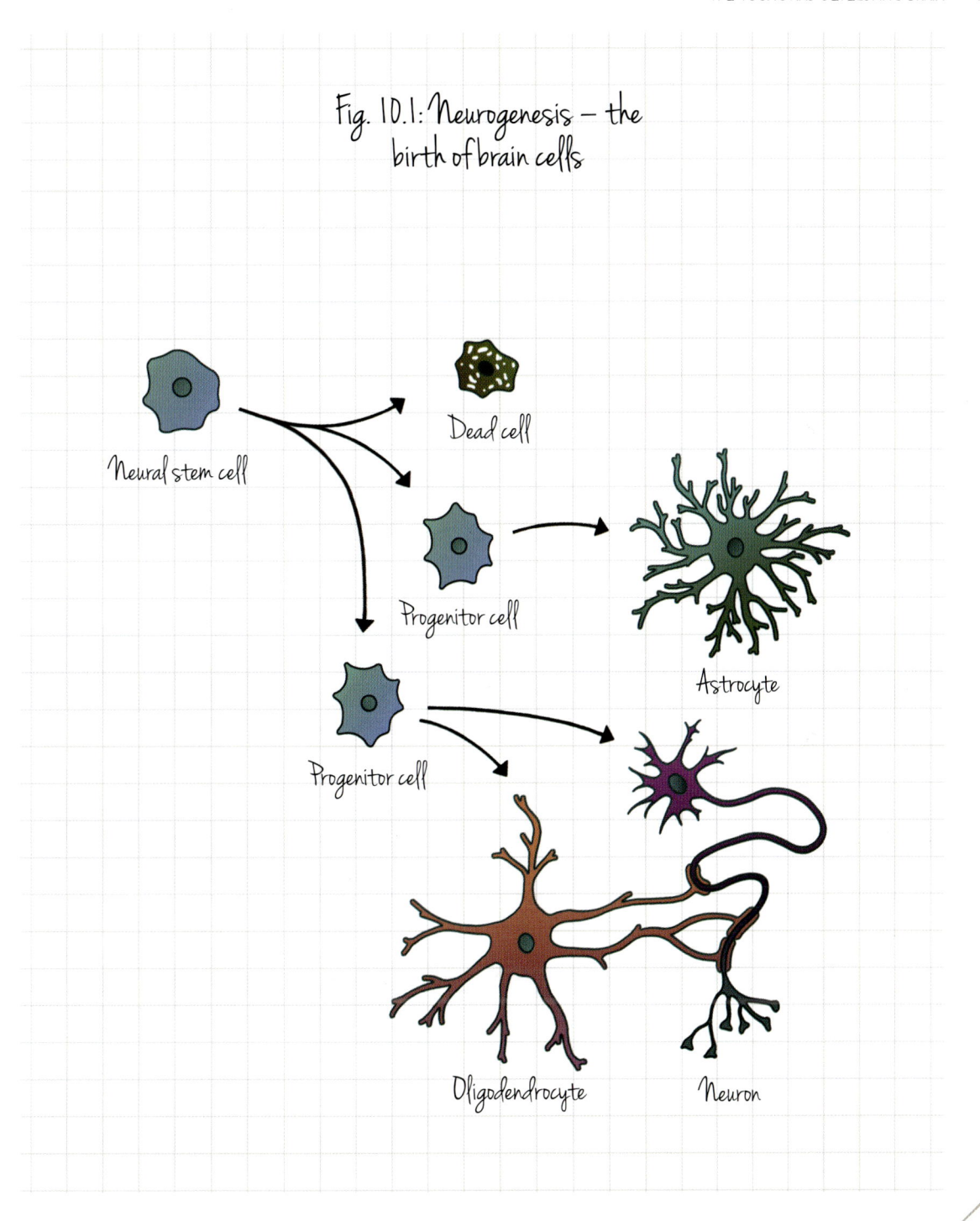

Fig. 10.1: Neurogenesis — the birth of brain cells

Neural stem cell

Dead cell

Progenitor cell

Astrocyte

Progenitor cell

Oligodendrocyte

Neuron

122　THE YOUNG AND DEVELOPING BRAIN

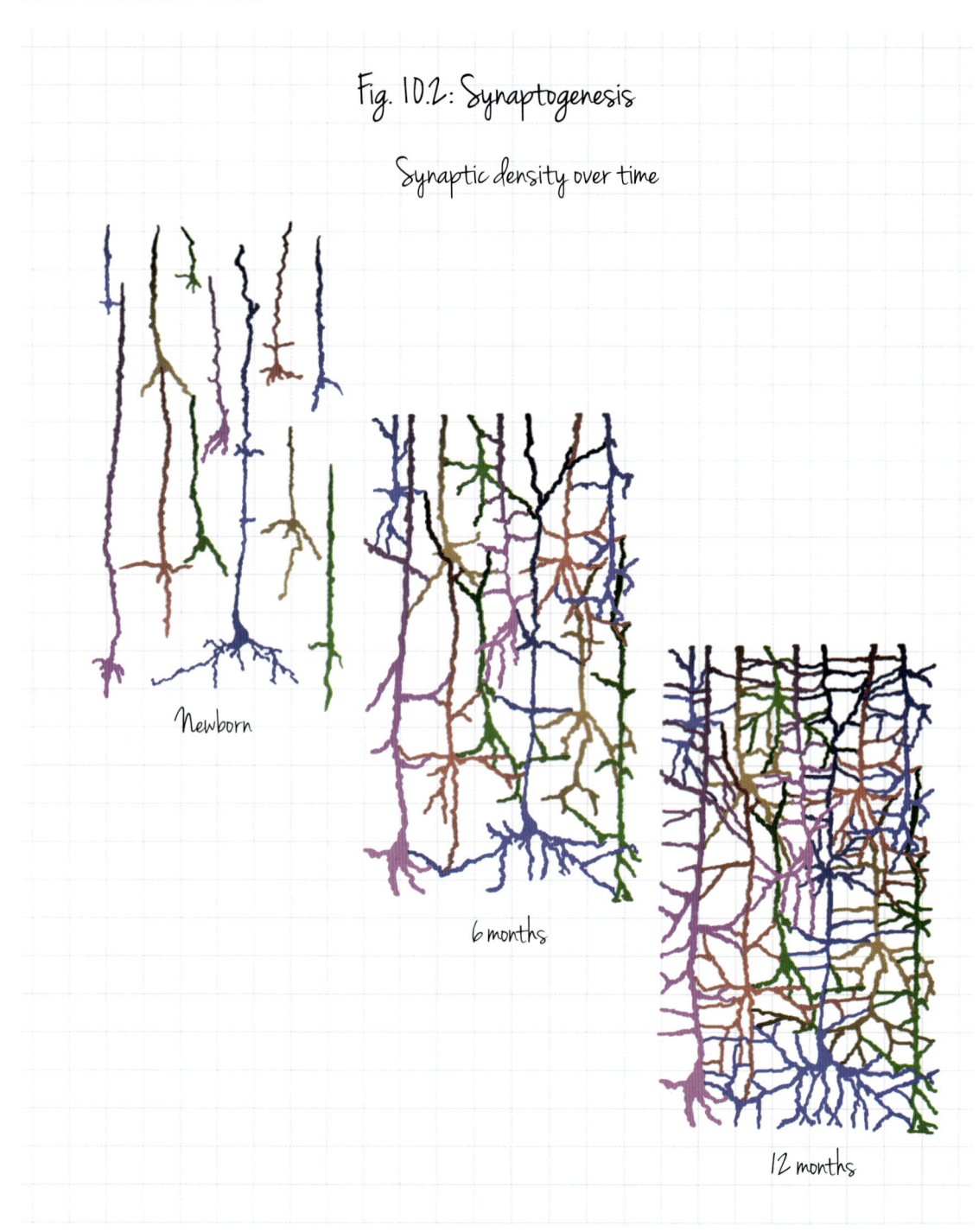

Fig. 10.2: Synaptogenesis

Synaptic density over time

Newborn

6 months

12 months

able to grow new neurons in the cerebellum. However, by and large it seems that there is little, if any, neurogenesis in any other parts of the brain after birth. This is why brain injuries, strokes and degenerative diseases such as Alzheimer's and Parkinson's, or indeed spinal injuries, are so devastating – on the whole, we just cannot replace lost nerve cells.

There is good reason for a general lack of neurogenesis in the human brain: the structure becomes so complex and finely tuned that simply replacing lost cells with new ones would entirely disrupt the system. So, rather than producing new cells, the bulk of the brain-building process occurs through synaptogenesis – the development of new synaptic connections; and myelination – the growth of an insulating layer that wraps around axons to allow them to communicate further and up to one hundred times faster. These changes are relatively easy to observe with modern techniques. Myelin is white, so it shows up as *white matter* on MRI and CT scans (*see* pages 15–16). This contrasts with cell bodies and synapses, which show up as *grey matter*. Numbers of synaptic connections can also be observed by taking very thin slices of the brain and examining dendrites, the branch-like extensions that allow neurons to connect with each other. Together, synaptogenesis and myelination allow any individual brain to develop new, stronger and faster connections in areas of the brain that are being regularly stimulated.

The last, but equally important, part of the brain-building process is pruning – getting rid of neurons and synapses that are not needed. The formal term for this deliberate loss of neurons is *apoptosis*; this is distinct from *necrosis*, which is the term used to refer to cells that die as a result of infection or trauma. It may seem counter-intuitive to kill off perfectly healthy cells, especially when we know that there are very few new ones produced after birth. However, just as a rose bush or a tree will grow better and produce stronger flowers after pruning, so the brain functions more efficiently if the un-used neurons are stripped away. In fact, some scientists have hypothesized that autism and schizophrenia may be the consequence of inefficient pruning of neurons and synapses. There is still some controversy around this theory, but certainly

Above: In the same way that pruning is necessary for the healthy growth of a plant, it is an essential part of the brain-building process.

increasing evidence that the neurodevelopmental process follows a different trajectory in these individuals and results in atypical connectivity between parts of the brain.

Although the human brain is continually adapting and growing throughout life, the biggest bursts of synaptogenesis, myelination and pruning occur during infancy, adolescence and old age. Given that these are such active phases of neurodevelopment, it is perhaps not surprising that they are also periods of particular vulnerability for the developing brain. Many conditions are most likely to first present themselves at one of these times – for example, autism, ADHD, Tourette's syndrome and dyslexia tend to be diagnosed during infancy and early childhood; schizophrenia, depression, and anorexia nervosa often make their first appearance during adolescence; Parkinson's disease and various forms of dementia most often occur during older age. Successful neurodevelopment depends on the balance of growth and pruning in different parts of the brain at these crucial stages. These processes are in turn driven by a complex interplay between environmental factors and pre-determined genetic codes, programmed to come into play at different times in life.

THE INFANT BRAIN

The first signs of brain development begin just three weeks after conception and, by the end of the embryonic period, eight weeks into gestation, all of the rudimentary brain structures are already in place. Neurons themselves start appearing from the sixth week, and through the course of prenatal development they gradually migrate to different brain areas and form connections that establish the basic neural pathways. By the time a full-term baby is born, all the major pathways are in place. These are like the motorways in the brain, carrying signals from one major region to another.

While there are few new neurons added to the brain after this point, it continues to grow and develop rapidly, quadrupling in size during the pre-school period and reaching approximately 90 per cent of the adult volume by as young as 6 years. Much of this growth is the result of synaptogenesis, as the young neurons form extensive connections with their neighbours. In fact, this growth of synapses is so dramatic in the first year of life that synaptic density in a 1-year-old child greatly exceeds that of an adult. Alongside this there is also a rapid increase in white matter, as the neurons become increasingly myelinated, making them faster and more efficient. Over recent years, research has been able to track these different elements of brain growth across childhood and has revealed two important things.

Below: By the time a baby is born, all the major neuronal pathways are in place.

Fig. 10.3: Stages of brain development in an infant

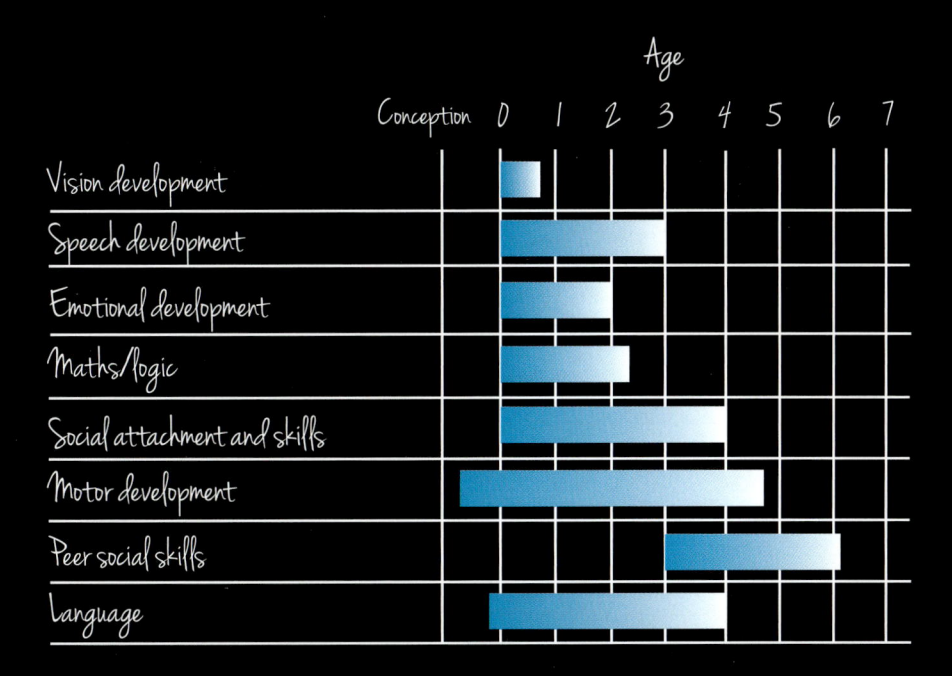

Skill	Age range
Vision development	0–0.5
Speech development	0–3
Emotional development	0–1.5
Maths/logic	0–2.5
Social attachment and skills	0–3.5
Motor development	Conception–5
Peer social skills	3–6.5
Language	0–3

The first of these is that every exuberant burst of growth is followed by a significant period of competitive pruning, which is driven by individual experience. This was illustrated brilliantly in research by the French neuroscientist Olivier Pascalis and his colleagues at the University of Sheffield. They showed the faces of monkeys to six month- and nine month-old babies and also to adults, and then measured their respective abilities to recognize them. The youngest infants greatly outperformed the adults, but by the age of nine months the monkey-recognition skill had been lost. Since this coincides with the pruning period in this part of the brain, it has been offered up as evidence that we fine-tune our abilities according to the environment we find ourselves in. Presumably the character Mowgli from *The Jungle Book* would have pruned out different neurons, given the number of monkeys he played with!

The other significant thing about brain development is that each surge of connectivity and pruning occurs at different times for different regions. Sensory and motor functions are the first to mature, with all myelination complete by the time a child is four years of age. The process occurs a little later for areas involved in language and later still for those involved in strategic thinking and planning. All of this aligns well with the observed pattern of sensory, motor, language and emotional and cognitive development. For example, the so-called Terrible Twos (see page 23) reflect a stage at which children's brains are sufficiently developed for them to know what they want and to be able to express that using behaviour, yet their language does not allow them to communicate very clearly. This is because the brain region involved in inhibiting and regulating behaviour – the prefrontal cortex – is still a long way from maturity.

Above: In the 1960s David Hubel and Torsten Wiesel carried out a series of important studies on domestic cats.

Interestingly, as we will see in the next chapter, this same region of the brain is one of the first structures to be affected by age, which may partly explain why some older people can be a little forthright in expressing their views.

ENRICHED ENVIRONMENTS AND CRITICAL PERIODS

One of the big debates taking place in educational policy over the last decade has been at what age children should start school. This varies widely across Europe, with children in Northern Ireland starting compulsory school at 4, in England and Cyprus at 5, in Italy and the Netherlands at 6, and in Poland, Sweden and Finland not until the age of 7. Interestingly, the later start does not seem to have any negative impact on how much individuals achieve by the time they leave school – and some have even argued that it might be beneficial to start school at a later age. There are many factors influencing this decision, but educational policy makers have leaned heavily on two key pieces of neuroscientific evidence to support the case for an earlier start: *critical periods* and *enriched environments*.

Back in the 1960s, there was a series of important studies carried out on domestic cats by a duo of neurophysiologists named David Hubel and Torsten Wiesel. They deprived kittens of one of their eyes by sewing it up soon after birth, not re-opening it again until the cat reached adulthood (around six months). Although the eye that had been blocked still appeared to function normally, they found that the brain cells that would normally respond to that eye were unresponsive and had shrunk. In contrast, the area of cells that were allocated to input from the unaffected eye was bigger than normal. In other words, the visual cortex adapted itself so that the parts that were not being used were made available to the other eye. However, the really crucial finding was that this did not happen if the experiment was repeated in adult cats. In fact, Hubel and Wiesel were able to show that this effect occurred only if sight was blocked off sometime in the first three months of life.

We now know that a similar thing happens in both monkeys and humans, and this has led to the idea of *critical periods* – that relevant brain functions will developonly if there is relevant input within a given time frame. Other studies have shown similar effects in other modalities – for example, an infant who is not exposed to a broad range of flavours between the ages of six and nine months may become a fussy eater. A similar case was made for language, but here the findings were far less conclusive; while it seems that it may take more effort to acquire language after the end of the proposed critical period (this varies between the age of 5 years and puberty), it is certainly not impossible. For this reason, most people now prefer to use the term *sensitive periods* to refer to the ages during which the relevant areas of the brain are at their most *plastic*.

Plasticity seems to be very dependent, not only on the type, but also the amount of stimulation the developing brain receives. In 1987, the American neuroscientist William T. Greenough carried out an experiment, which is often referenced by policy makers and "smart toy" makers alike. Building on earlier work by Mark Rosenzweig (*see page 71*), he reared genetically identical rats in three different environments – impoverished (bare necessities), simple (basic but not exciting) and enriched (a big space, lots of toys and other rats to interact with). He looked very closely at the way neurons had developed and found that the neurons of rats reared in the enriched environments had more elaborate branching and significantly more synapses.

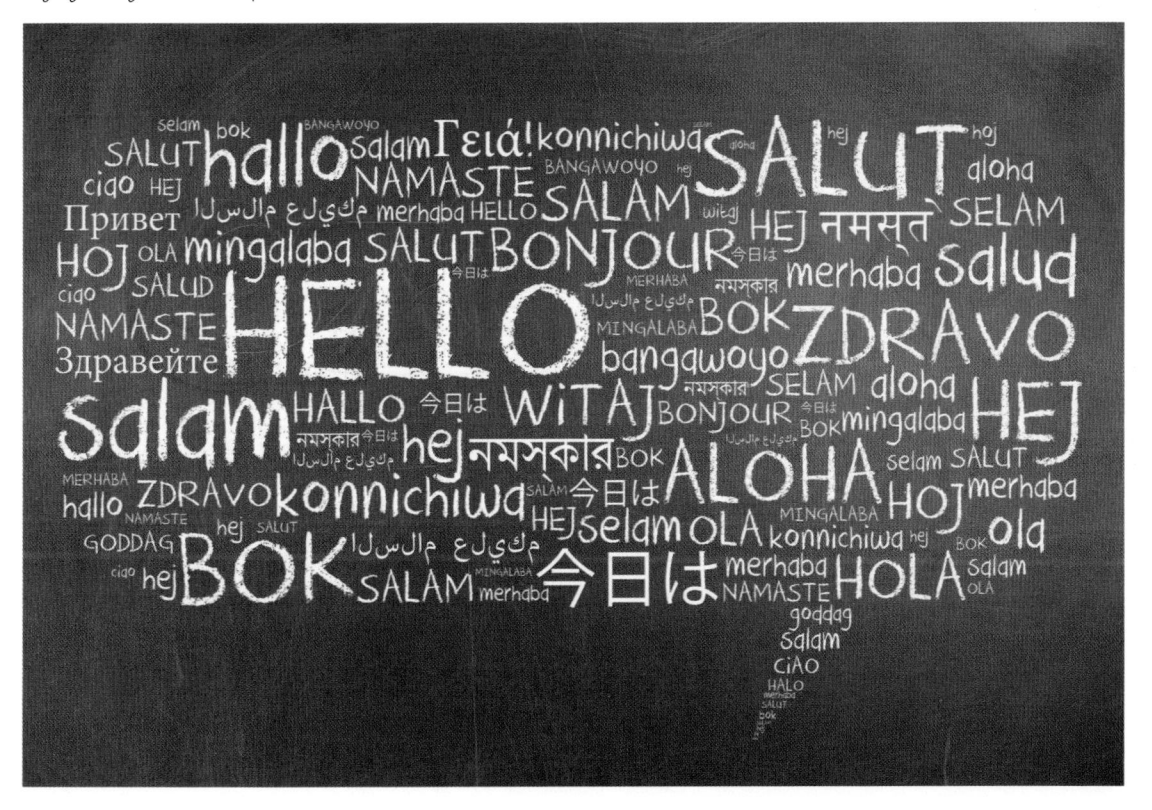

Below: The brain is highly receptive to learning any language during the "sensitive period".

ROOM FOR INTERPRETATION

While this is clearly an impressive and very significant finding, some people now believe these results have been used to tell a misleading story. The basic implication of Greenough's experiment is that simply by making an environment more stimulating, we can build bigger brains. This is certainly in line with what Hubel and Wiesel found, and also with other evidence of plasticity. However, we have to remember that we cannot assume that anything observed in the brains of rats will directly reflect what happens in human brains. But more to the point, Greenough's *enriched environments* are actually very close to what a rat would naturally encounter in normal life; so, rather than these living conditions being conducive to extra brain growth, could it be that the other two environments were detrimental to normal development? It may be that our children do not need enriched environments at all – indeed, there are many experts who suggest that the popular habit of "hot-house" schooling may have a negative impact – but rather that we need to guard against impoverished environments.

This theory has been strongly supported by Michael Rutter's very detailed investigation of a group of extremely deprived orphans originating from the regime of the deposed Romanian dictator Nicolae Ceaușescu. Rutter examined the brains and functioning of a number of Romanian children who were brought over to the UK and adopted by families. He found that those who had been taken away from extreme deprivation at an age of around six months went on to develop brains that were equivalent to those of their adopted British siblings. However, the older the Romanian orphans were when they were brought to the UK, the less likely this was to be the case. Their brains tended to be smaller and they showed less intellectual and emotional development. This seems to support the idea that it is not that children

Below: The number of connections in a rat's brain is highly dependent on their environment.

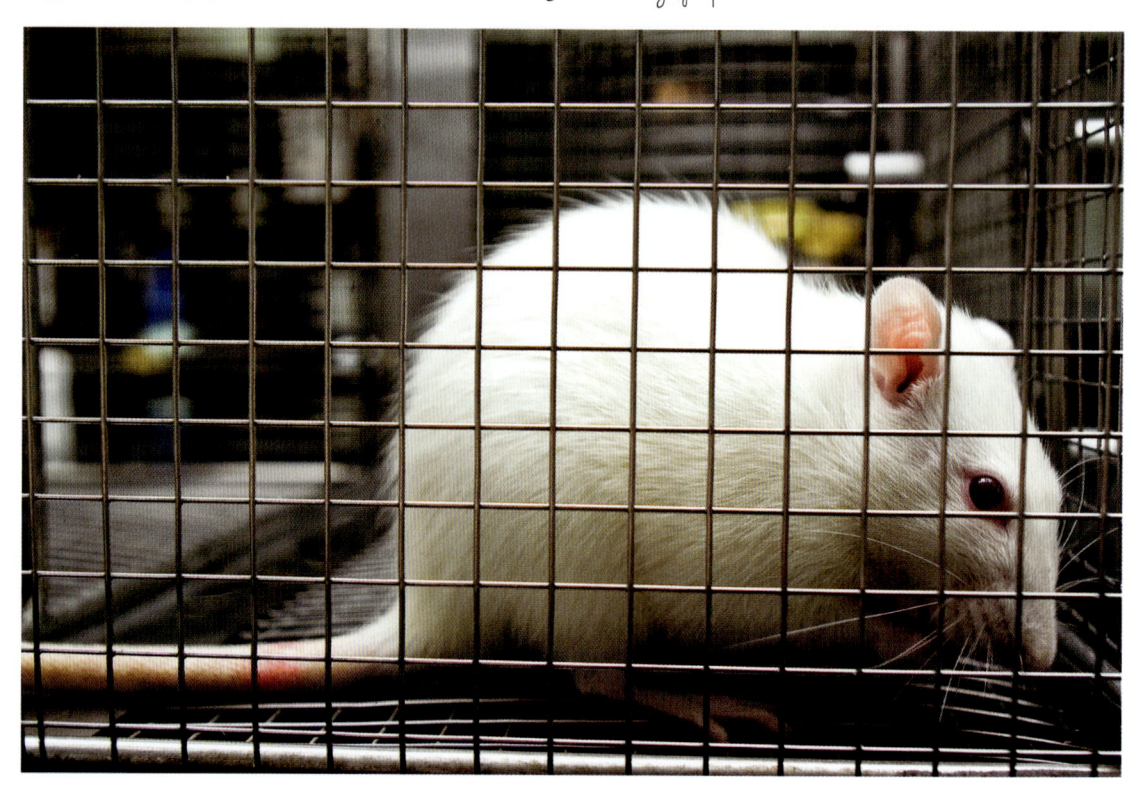

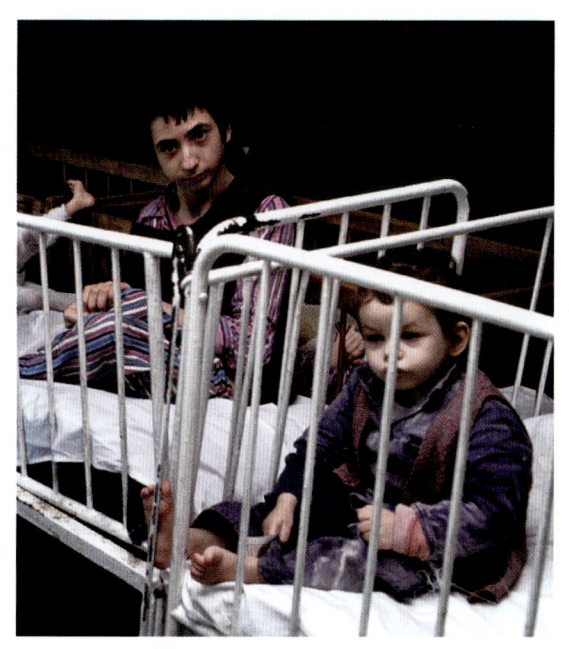

Above: Romanian children, known as Ceausescu's orphans, in a state-run orphanage.

need excess stimulation, but rather that they simply need to be fed, loved and allowed to explore their surroundings as they grow up.

THE TEENAGE BRAIN

The dramatic changes that happen to children in the first few years of life are challenging and exciting for any new parent; however, many are unprepared for the next major wave of change – when puberty strikes. Although many adults frequently reminisce fondly about their own teenage years, this is often a tricky period in a person's life. The American novelist Stephen King is quoted as saying, *I hated high school. I don't trust anybody who looks back on the years from fourteen to eighteen with any enjoyment. If you liked being a teenager, there's something wrong with you.*

King is not alone: back in 1904, adolescence was described by the pioneering American psychologist G. Stanley Hall as a period of *storm and stress*, and this view has stuck to some extent. Indeed, this phase of life is widely seen by neuroscientists as a window of vulnerability, with between a third and a half of all adolescents reporting significantly depressed mood or anxiety at any one time. However, it has more recently been described as a window of opportunity, because the growing brain is also particularly receptive to positive environmental influences. It is certainly a very complex period of human development, characterized by huge biological, psychological and social change. As one 17-year-old put it, *It's like you're not a child anymore but you're not an adult either; everything is changing.*

Sarah-Jayne Blakemore, a British professor and leading expert on the teenage brain, defines adolescence as the period between the onset of puberty and the time at which a person has a stable, independent role in society. In terms of the brain, there is considerable re-structuring during these years. The changes can nowadays be easily measured by looking at the ratio of white matter (myelinated axons) to grey matter (cell bodies and synapses). MRIs and post-mortems (see pages 15–16) both indicate that all the way through childhood and adolescence there is an increase in white matter, which suggests that neurons are developing longer, stronger and faster connections. However, as soon as an individual hits puberty, there is a big wave of synaptogenesis, which as always is followed by a phase of pruning. This happens at a different time and pace for different parts of the brain, which may explain some of the typical teenage social, cognitive and emotional characteristics.

In the prefrontal cortex, which is involved in planning, control and empathy, the number of synaptic connections seems to be at its peak in early adolescence (age 11 for girls and 12 for boys). After this, the amount of white matter continues to grow, but there is a sharp decline in grey matter, indicating pruning. This gradual reversal of the white/grey matter ratio continues until the late teens or early twenties and is in line with general improvements in cognitive tasks, such as working memory and processing speed. In contrast to this, some of the subcortical areas – those that are involved with emotion processing and the reward pathway – reach maturity rather earlier. Some researchers have proposed that this mismatch in development might underlie the increased risks that are often taken during teenage years. The idea is that the subcortical reward pathways become stronger but the prefrontal inhibitory mechanism lags behind, so that it is not always powerful enough to keep behaviour in check. Professor Blakemore likens this to driving a very fast and exciting car that has extremely poor brakes!

HORMONAL CHANGES

Of course, another big biological change that occurs during this time is a shift in hormonal balances. The sex hormones – testosterone and oestrogen, for example – have powerful influences on social, emotional, sexual and cognitive functioning. However, there are also changes to the levels of other hormones – in particular cortisol and melatonin – which have implications for stress, attachment and sleep. Recent studies have shown that the switch in sleeping habits that often occurs during teenagedom – which can be quite sudden and marked – is probably a result of alterations to the regulation of these particular brain chemicals. Therefore, a teenager who does not want to go to bed and then cannot get up the morning after might well be responding to an alteration in their natural biological rhythms. In fact, this science has so much weight now that there are pilot schemes taking place across the UK to investigate whether a later start to the day in secondary schools improves academic performance.

Optimizing the learning environment for teenagers is important, because this is a time when the brain is particularly ready to develop new skills and knowledge – the *sensitive period* for intellectual growth. However, these rapid brain changes also mean that individuals have a greater susceptibility to stressful experiences, especially when it comes to relationships, and a greater vulnerability to the neurotoxic effects of substances such as alcohol and cannabis. In some cases these substances can cause long-lasting changes to brain function, possibly even leading to psychosis. We must remember, though, that this is a time when individuals grow through exploration, and this almost inevitably involves taking emotional and behavioural risks. The transition from child to adult will always be challenging and requires careful balance between exploration and risk. Getting to grips with the changes that occur in the brain during this period is an important part of ensuring appropriate and useful support to individuals journeying through this window of both vulnerability and opportunity.

Below: A teenager who does not want to go to bed and finds it hard to get up in the morning may be responding to natural biological rhythms.

MYTH BUSTER: BRAIN GYM RE-PATTERNS THE BRAIN

Brain Gym is a set of 26 physical exercises that are said to *re-pattern the brain* and improve learning skills, self-esteem and behaviour in the classroom.

The Brain Gym programme was first developed in the 1980s by a couple of Californian "professional educators", Paul and Gail Dennison. It ticked along quietly for twenty years or so until the early 2000s, when it hit the headlines, became big business and evolved into a widely used, government-supported intervention for schools. Nowadays, Brain Gym is purportedly being used in schools in as many as eighty-seven countries.

In order to become a fully qualified Brain Gym practitioner, one must undertake extensive training and pay fees in the region of £4,000. Exercises include wiggling the ears in such a way that it *stimulates reticular formation and tunes out distracting and irrelevant sounds*, and putting pressure on the *brain buttons* in order to *re-establish brain organization for reading and writing*.

In 2006 there was a big backlash from the scientific community, which declared that Brain Gym was pseudoscientific nonsense, with the British academic and science writer Ben Goldacre describing it as being *barkingly out to lunch*.

While it is certainly true that exercise itself can be a useful part of the school day – it stimulates the cardiovascular system and raises alertness – there has so far been no convincing empirical evidence that this particular programme has any impact on learning at all. Even worse, it teaches children neurobiological explanations that have no basis in real science whatsoever. The idea that it is possible to access the brain by massaging the clavicle with one hand and rubbing the stomach with the other is, as Goldacre put it, like believing that you can influence your central heating system by rubbing the wall in front of the pipes.

Appealing though it may sound to be able to re-map the learning parts of the brain using a few simple exercises, the strong message from neuroscientists is not to be seduced by the pseudo-science of Brain Gym.

11. THE TWILIGHT YEARS

HAVE YOU EVER WALKED INTO A ROOM TO FETCH SOMETHING, ONLY TO FORGET WHAT IT IS YOU WENT IN THERE FOR? ALMOST CERTAINLY!

Which of us has not struggled to remember the name of a colleague, film star or politician, only to have it pop into our mind a few hours later when we are no longer trying? We are all prone to memory lapses from time to time, but these so-called *senior moments* tend to increase as we get older.

On the other hand, there are some mental abilities that continue to improve with age and there are also people who seem particularly resilient to the ravages of time. Research from the neuroscientist Professor Lorraine Tyler at the University of Cambridge has described 80-year-olds who perform as well as much younger people on memory tasks. She also found that their brains actually looked younger. So, can we do anything to stave off the effects of age? This chapter will explain how the brain ages, what effects that might have on the way we think and feel and, most importantly, what we can do to keep our brains and minds healthy.

UNDERSTANDING WHAT HAPPENS TO THE BRAIN AS WE GET OLDER

In 2009, at the age of 95, having survived open heart surgery and two forms of cancer, American athlete Leland McPhie set a world record in the long jump for veterans, which currently still stands. A lifelong lover of track and field events, Leland continued competing until he was 100 years old, but he finally died of an E-coli infection in 2015 at the grand old age of 101. Of course, Leland was something of an exception, but centenarians are the world's fastest growing age group; very rough estimates suggest that there are around 450,000 of them currently alive across the globe. As a species we are certainly living longer, but while some people retain a good level of mental and physical fitness all the way through these later years, others are not so lucky. Fear of dementia is a big issue for people heading towards their twilight years, with a recent study showing that this is by far the biggest health concern in those over the age of 55.

Brain scans and post-mortems have revealed that there is a huge variability in brain structure, not just between individuals, but also within the brains of the same individuals. Some areas of the brain seem to be particularly vulnerable to the effects of ageing – for example, the limbic system and prefrontal cortex – whereas others are quite resilient. Therefore, while some things may become more difficult, others will not. This means that for those that follow a relatively normal ageing trajectory, there will be some cognitive abilities that continue to grow into very old age.

The bad news is that, from a biological point of view, the ageing process begins frighteningly early – as early as our twenties! The good news is that although we lose brain cells very early on in our lives, this is not always a bad thing. In fact, as we saw in Chapter 10, this is one of the ways in which our brains develop and become more efficient. Even where the losses might be detrimental, the brain is remarkably adept at compensating for these changes. So, while some brain cells become older and less efficient – and will eventually be lost altogether – the stronger cells will be more likely to survive. Additionally, as we get older, these

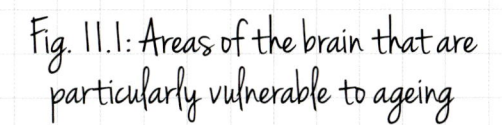

Fig. 11.1: Areas of the brain that are particularly vulnerable to ageing

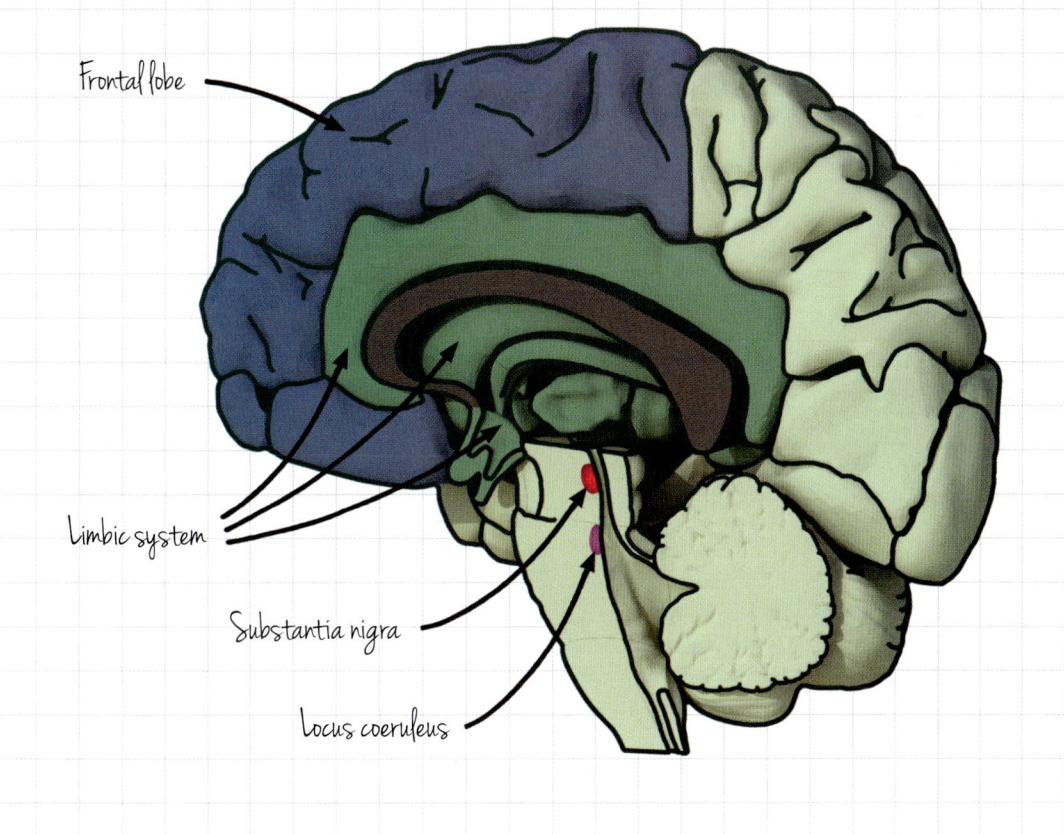

Frontal lobe

Limbic system

Substantia nigra

Locus coeruleus

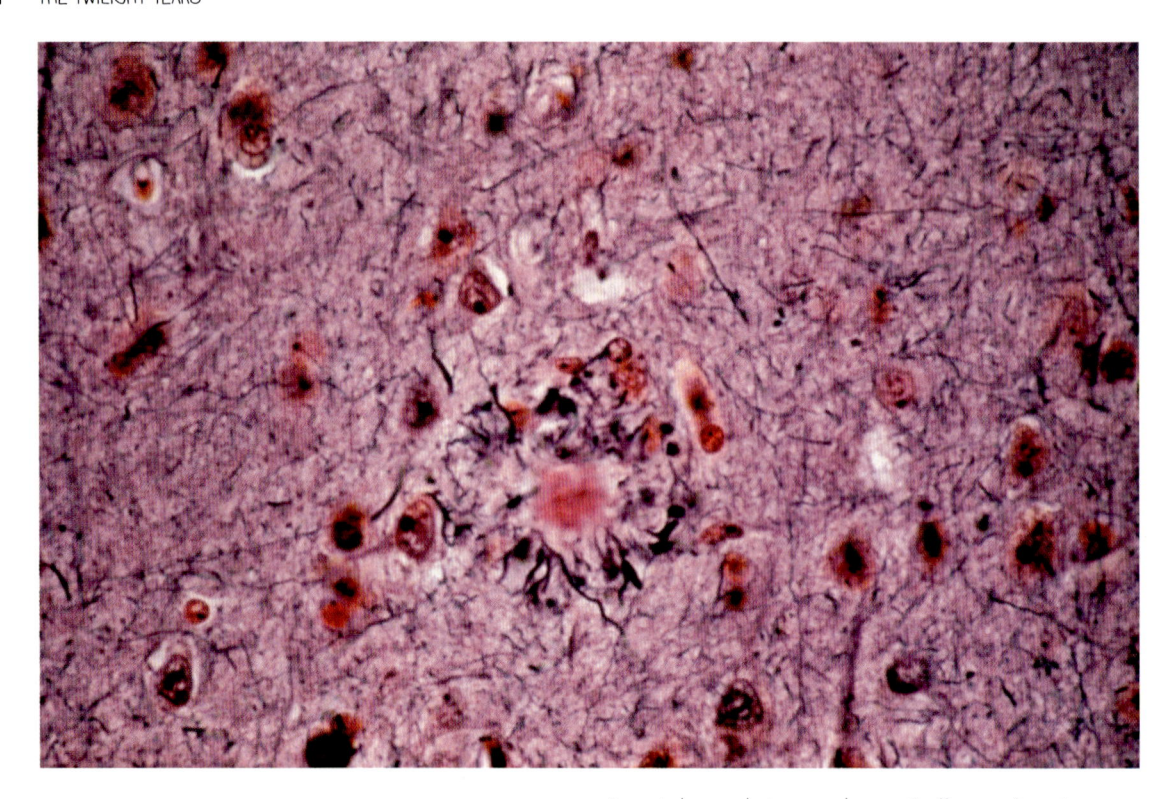

Above: A photograph depicting the neurofibrillary tangles and amyloid plaques that increase in number as we age.

cells continue to grow and build new connections. Cellular research has shown that brains show plasticity well into the eighth decade.

BRAIN DAMAGE

Nevertheless, as we creep toward retirement age, those brain cells that survive are more likely to become damaged. There are two important hallmarks of the ageing brain – *neurofibrillary tangles* and *amyloid plaques*. Neurofibrillary tangles are clumps of a specific protein called *tau*, which build up inside some of the neurons, particularly those in the hippocampus. Higher numbers of these have been linked with poorer cognitive performance and very high levels are a specific marker of Alzheimer's disease (*see page 39*). It has generally been believed that these tangles interfere with signalling within the neuron and may increase the likelihood of cell death. However, some have recently suggested that the opposite may in fact be the case – that they may be more prevalent in damaged brains, because they are a protective mechanism, like an army sent in to rescue the good cells. Amyloid plaques are lumps of beta amyloid protein that build up **between** the neurons and are also thought to impede synaptic communication. Again, these are far more prevalent in the brains of people with Alzheimer's disease and are seen as an important marker.

One very important way of keeping the brain fit and well is to maintain a healthy cardiovascular system, because this plays an essential role in carrying nutrients to the brain and removing damaging toxins. For this reason, anything that reduces or disrupts blood circulation will impact on brain health to some extent. However, sometimes the effects can be quite dramatic – for example, when someone has a heart attack or a stroke. Strokes – often described as *cardiovascular accidents* – refer to situations in which an area of the brain is deprived of oxygen, either due to a blood clot or because of a blood vessel rupture that causes bleeding. Either way, a stroke usually leads to the permanent loss of brain cells. These events may cause significant, possibly lethal, damage to large areas of the brain, or they may be so minor that they pass unnoticed and result in very few effects. However, all in all, there is a high correlation between a healthy heart and a healthy head.

HOW DO BRAIN CHANGES AFFECT MEMORY AND OTHER MENTAL FUNCTIONS?

Even the fittest brains will show some effects of age, so what are the changes we can expect to see in a typical healthy individual, as they get older? Research consistently shows us that the cognitive functions that are most likely to be affected by age are: (1) overall thinking speed – for example, having to make quick decisions, process directions that are given, or carry out mental arithmetic; (2) "executive functions" – this includes things such as planning, thinking ahead, using complex strategies and being flexible (for example, having to deal with changing plans); and (3) memory – in particular struggling with people's names, being able to find the right word, or remembering to post a letter. We will look at each of these in more detail as well as considering ways in which the brain continues to improve.

THE EFFECTS OF AGEING

A predictable and pervasive consequence of ageing is a general "slowing down" in almost everything in life. When it comes to the brain, we know that from our twenties onward there is a very small but continuous loss of white matter, meaning that numbers of myelinated axons are decreasing. These are like the motorways carrying fast information around the brain, so the overall effect is a very gradual reduction in thinking and processing speed. This may not become noticeable until people reach middle or even later life, especially since it is balanced up by the fact that people generally become more knowledgeable and practised at doing things. However, there are clearly measureable changes in processing speed by the time people are in their sixties and seventies and some scientists believe that this is the main explanation for a decline in memory and other cognitive tasks. If this is the case, then simply giving people a bit more time to accomplish a task would make all the difference, and studies suggest that to a large extent this is true.

An alternative, although not necessarily contradictory, view is that ageing causes more specific changes to the brain – that only some regions become less efficient, while the rest continues to function well. The prefrontal cortex and hippocampus seem to be particularly susceptible to the effects of age, which fits well with the cognitive changes listed above. The frontal lobes are often described as last in and first out in terms of development, so there are some cognitive parallels between the younger child and the older adult. As individuals become older, they tend to become a little less inhibited and may take less time to stop and think before making decisions or expressing their thoughts. People may also become more set in their ways or may get caught up in talking about a particular topic and find it hard to move on. Planning and thinking ahead also becomes more difficult. Of course, these effects may be quite subtle and do not affect everybody.

A failing memory is probably the most common cognitive complaint from older individuals. People are

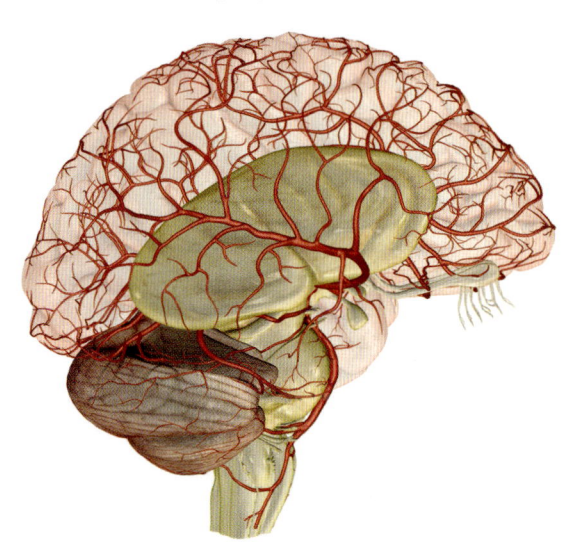

Above: Maintaining a healthy cardiovascular system is essential in keeping the brain fit.

often not as impaired as they think, but there are some generalized changes most can expect. Whatever age we are, it is typically more difficult to remember information that has no specific meaning – for example, names or strings of unrelated numbers. Memory for this type of information, especially names of friends or celebrities, is one of the first cognitive difficulties that people notice, even as early as in their thirties.

Another problem that is often experienced by people in middle adulthood is that words do not come to mind so easily, but this gets significantly worse as people get into old age, and little things they feel they really ought to know can seem completely out of reach. These feelings are frustrating but incredibly common, and often the information will come back to mind a bit later, usually when the person stops consciously trying and gets on with something else. Age also takes its toll on *prospective memory* – remembering to do things at a particular point in the future, which may mean anything from turning up to an appointment, to buying some stamps when out shopping. Again, this memory problem is not limited to being older but it does worsen in later life and can be particularly worrying if it means forgetting to take medication or turn off the gas fire.

It is very common for people to notice that memories of things that happened to them during their schooldays or early adulthood are far clearer than their memory of yesterday. Research certainly shows that most healthy older

Below: (Above) Three MRI scans of the brain of a normal 22-year-old man. (Below) Three scans of the brain of a 96-year-old woman. The older brain has less grey matter and larger internal spaces.

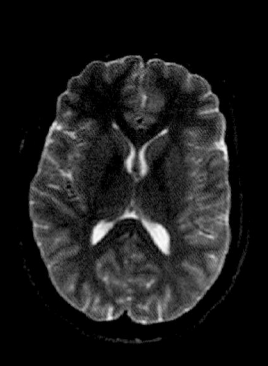

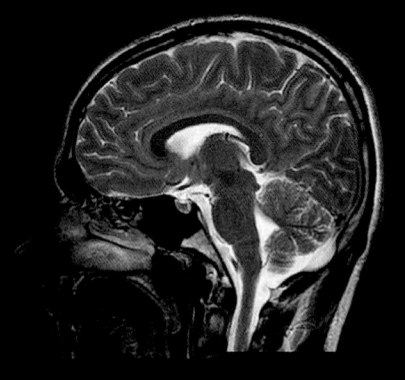

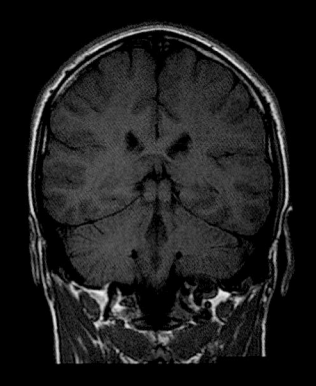

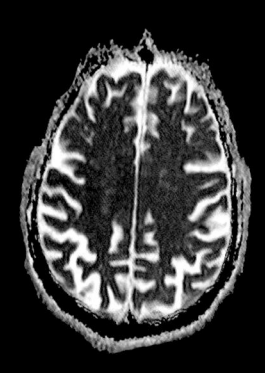

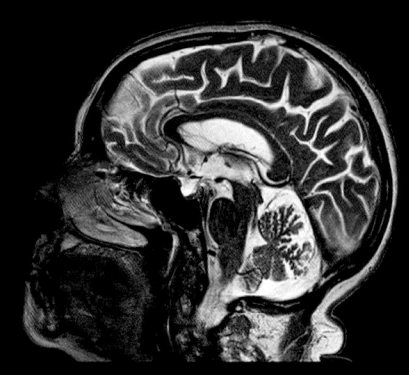

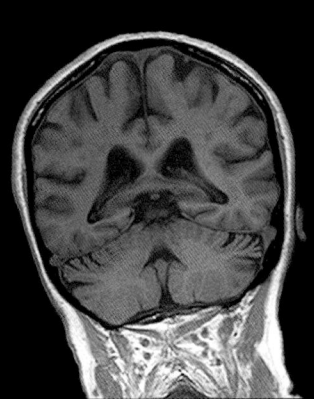

individuals have a very good memory for their twenties and thirties, but when they are asked to recall information they have just learned, they perform quite poorly compared to their younger counterparts. Interestingly, the difference between older and younger people becomes much smaller if some cues or aide memoirs are given. Most studies demonstrate that if people are simply asked to recognize rather than recall (for example, on a multiple choice test), older people do no worse than anyone else. So, although it may seem that information has simply vanished, it may in fact just be temporarily out of reach and relatively easy to retrieve, given the right cue and sufficient time.

In contrast to all of these cognitive losses, vocabulary, knowledge and wisdom generally continue to grow, so an older person can be a very useful addition to any quiz team! Everyday problem-solving also tends to be unaffected by age and can even continue to improve, except where there is significant time pressure. There are also some aspects of memory that seem to be completely unaffected by age: in particular, anything that becomes a well established routine or habit. Very short-term memory tasks, such as temporarily holding a simple piece of information in mind (e.g. a short phone number), also seem to be largely unaffected, so long as there are no distractions.

AGE-PROTECTING THE BRAIN AND "POSITIVE AGEING"

Despite the endless and convincing promises that are so prevalent across the media and advertising billboards, there are no magic solutions to ageing and no elixir of youth. The biological changes we have discussed in this chapter are inevitable for those of us lucky enough to have a long life. But as we have seen, there are vast differences between how well people age and, although some of this will be genetically determined, environmental and lifestyle factors are hugely relevant. Some factors can actively protect the brain by boosting the growth of neurons and synapses or reducing cell loss, but it is also possible to learn specific tricks for maximizing those abilities that may be failing.

Right: One of the affects of ageing is a general slowing down in all aspects of life.

Above: Regular walking is enough to make a significant difference to brain health.

Without question, the most effective age-protecting step anyone can take is physical activity. There is now very convincing evidence that people who exercise and are regularly active have healthier brains – more brain cells and more synaptic connections – and consequently perform better on tests of memory, planning and problem-solving. Quite simply, a healthy body means a healthy brain and mind. It is not yet clear what the optimum amount or type of exercise is, but many studies have shown that even regular walking is enough to make a significant difference. And importantly, it is never too early or too late to start. Exercise seems to directly impact on neuron and synaptic growth, but it is also an excellent way of looking after the cardiovascular system, so that the brain is properly nourished and detoxified.

Two factors of ageing that can easily be overlooked are diminished eyesight and hearing. Both these key senses become less efficient with age. This has serious implications for those who suffer from a decline in these faculties, because it is quite easy to assume that someone has forgotten something when they might not have seen or heard it properly in the first place. You have only got to imagine wandering around with cotton wool in your ears and wearing some blurred glasses in order to appreciate how much more taxing and stressful the world would be if you could not see it or hear it properly. Under these conditions, information is not perceived in as much detail, which means that it is not processed as deeply. In addition, mental processing is far more effortful, which can lead to a sense of being overwhelmed by an excessive cognitive load. It may seem obvious, but ensuring that an individual can see and hear comfortably makes a big difference to understanding the quality of their mental function.

THE BENEFITS OF A HEALTHY DIET

Brain health can be directly supported through a good diet, high in fresh fruits and vegetables, low in processed sugars and animal fats. There is a modern obsession with the concept of "superfoods". This can change as fads come and go, but at the time of writing the list currently contains berries, seeds and nuts, dark chocolate, fish and a Mediterranean diet – consisting of lots of vegetables, lean meat and olive oil. For all of these there is some research that suggests they may have some value in supporting brain function, but evidence is sometimes weak and may come primarily from animal studies. Nevertheless, each of these has a wide range of nutritional merits and will certainly contribute to a healthy balanced diet.

There has been a particular focus on blueberries, which contain high levels of chemicals called *polyphenols*, a type of antioxidant that seem to provide good neuro-protection. There is some sound evidence that these might protect – possibly even boost – cognitive function, but they would probably need to be taken in very large amounts to have any really significant effect. Overall though, antioxidants are a good way of protecting against cell loss. They are abundant in most fruits and vegetables, which is one good reason to aim for that well-publicized figure of a *five-a-day* minimum of these foods. There is also good reason to avoid foods that are high in processed sugar and to eat slow-release carbohydrates such as brown bread and brown rice. This is because the body becomes less efficient at controlling glucose levels as it gets older, which can be problematic, because high levels are toxic and low levels mean the brain may be starved of energy.

Over the last few years, the media has presented conflicting reports on alcohol use – so what does the science say about a regular glass of wine or a whisky nightcap? Excessive intake of alcohol is harmful to the brain at any stage in life, but more so at crucial neurodevelopmental phases – that is, in adolescence and old age. In an older person, not only are brain cells more vulnerable, but also the liver is less efficient, so alcohol and its toxic by-products hang around longer and cause more damage. However, there is research that has controversially suggested that a small regular amount of alcohol, in particular red wine, may have benefits for some people. For example, a large study from Harvard University in 2011 assessed the drinking habits and health of 14,000 women, and found that those who consumed between half and an ounce of alcohol daily had fewer effects of ageing than those who were non-drinkers or heavy drinkers. So far there is not a large study that supports this in men.

A HEALTHY MIND IS A HAPPY MIND

Psychological wellbeing is just as important and has a direct impact on physical health and cognitive functioning. As we saw in Chapter 4, a high level of stress is chemically toxic to the brain: excessive cortisol can damage cells, especially in the hippocampus. Anxiety can also affect other bodily functions, such as glucose regulation and blood pressure, which can directly impact on brain health. So, while it is inevitable that life will be challenging at times, it is important to avoid prolonged exposure to stress and to find a lifestyle that includes relaxation and positivity. One of the biggest

Below: Brain health can be supported through a healthy diet, high in fresh fruit and vegetables.

stressors is any major life adjustment, such as bereavement or moving house, both of which happen more frequently in old age. So it is important that changes are minimized where possible and always fully supported. Active relaxation – meditation, massage, listening to music – can all make a big difference and it is important to have good sleeping habits (*see* box in Chapter 4, page 63).

Finally, there are a number of strategies that can be put in place to support a failing memory or diminished cognitive function. It always helps to turn regular important events into a habit – having a checking routine before leaving the house or going to bed, looking at the calendar or diary at a set time every day, or habitually keeping things in the same place. Writing things down or using other cues or memory aids can also help – putting a letter by the front door if it needs to be posted, having a calendar by the phone, or a memory board on the wall. Other simple things can make a difference too, such as making sure that there are no distractions and giving yourself plenty of time when there is something important to remember or work out. Writing a diary can make a real difference for some people as well, especially if this is done regularly, just before bed. It is a good means of cementing important memories.

PATHOLOGICAL MEMORY PROBLEMS

When are memory problems or other cognitive difficulties something to worry about and not just a normal part of ageing? This is a tough question to answer, but some things are more likely than others to signal pathological conditions, such as dementia. These include: struggling to recognize people or places that should be very familiar; noticeable frequent forgetfulness over very short periods (a few minutes); having difficulty following a film or a story; sudden changes in personality; and other neurological signs such as experiencing tremors. However, sudden shocks, medication, illness and depression can all cause similar symptoms. Therefore, the only way to be absolutely

Below: Good sleep is important for memory and concentration.

THE TWILIGHT YEARS   141

Above: Keeping notes and using a diary can help us to remember things we need to do.

TOP TIP FOR POSITIVE AGEING

Did you know that your attitude to ageing might affect the rate at which you age physically and mentally? The eminent American psychologist Professor Becca Levy carried out a 39-year-long study, which evaluated 18- to 48-year-olds for their attitudes to older people. Her team found that those with negative attitudes were significantly more likely to have early heart attacks and strokes. In a separate study she also found that people who held negative stereotypes had more memory problems over time. The message is clear: a positive attitude toward older people may slow down the rate at which you age yourself!

sure whether there is a medical problem or not is for the individual concerned to have a detailed medical and neuropsychological assessment.

Terry Pratchett, the highly successful and prolific British author, died in March 2015, nearly eight years after he was diagnosed with early onset atypical Alzheimer's disease. He inspired many people with his determination to continue writing, his willingness to talk about his experiences and his quest for more funding to support those with dementia. None of us can know what is waiting for us in our old age, but science has shown us that we can be proactive in slowing down cell loss, building synaptic connections and reducing the likelihood of strokes and heart attacks. Probably the most important finding is that it is never too early or too late to make a difference. Looking after our brain is something we can all do from the cradle to the grave.

12. HOW THE BRAIN SPEAKS

WE LIVE IN A WORLD RICHLY STEEPED WITH LANGUAGE - WRITTEN TEXT, SPOKEN WORDS AND, OF COURSE, THE ENDLESS CHATTER OF OUR OWN INTERNAL VOICE.

It is estimated that there are well over 6,500 different languages in use across our planet, and the extraordinary thing is that, given enough early exposure, we are all capable of learning any one of these. Did you know that to a Samoan, *U* refers to a type of snail, while for someone in Brazil the same word describes someone who eats meat, and if you are Burmese it refers to a male over 45? Such is the remarkable flexibility and creativity of the human brain when it comes to communicating detail. Our capacity to absorb, produce and create meaningful language is a fundamental, unique and very precious skill. Anyone who has spent time with a pre-verbal toddler will know the limitations and frustrations of trying to communicate without words. But what is the evolutionary value of speech and how does the brain produce and make sense of it?

Below: Reserach suggests that babies begin absorbing language while still in the womb.

Above: Some theorists believe that tool use and other hand movements have an evolutionary link with language.

DEVELOPING LANGUAGE

No one knows exactly when spoken human language first appeared, and there is little in the way of solid evidence about how it emerged. However, many people believe that we can learn a lot about the evolutionary development of language by examining communication in our ancestors, the great apes. Evolutionary psychologist Katie Slocombe, for example, has developed a fascinating translation kit for the grunts and calls made by chimpanzees. By recording and playing back different noises to chimpanzees, she has shown that they use what she calls *rough grunts* to indicate the presence of food. Fascinatingly, Slocombe also found that the chimps alter the nature of their grunts to indicate to their chums how nice, or valuable, the food is – long, high-pitched grunts mean that the food they have found is good, whereas low-pitched, noisy grunts are used for less exciting food. Other researchers have suggested that animals may even have the equivalent of a regional accent – apparently ducks from London have a different quack from those in Cornwall, and primate gesturing systems may also vary according to the group they come from.

Another particularly interesting line of enquiry has been pursued by Gillian Forrester, a cognitive neuroscientist from the University of Westminster, who has spent a number of years investigating how gorillas interact; she studies precisely what noises, actions and facial expressions they use towards each other. In particular, Forrester is interested in tool-use activities that are common to both humans and great apes, and which she believes may be a precursor to human language. She argues that the sequences of hand actions used by great apes may have some structural parallels with the way that human sentences are formed. Indeed, there is good evidence for the hypothesis that language may have developed from gesture, not least due to the fact that the hands and mouth share some of the same neural mechanisms. Forrester has also carried out a series of related investigations with young children; one of her recent experiments suggests that the inherent link between hand movements and language may explain the intriguing observation that young children

Above: Language allows us to share our memories and feelings, which forms the basis of relationships.

often stick their tongues out when they are concentrating on particular tasks.

While many animals are quite able to communicate their current emotional state and can sometimes warn of danger or indicate the presence of food, we do not currently have the knowledge to investigate whether they have the complexity of language that we do. However, as a concept this seems unlikely, and most scientists tend to believe that human language is unique. Our highly evolved communication system enables us to share our memories of the past; make plans for the future; develop elaborate relationships; preserve our thoughts, feelings and transactions; and engage in complex negotiations. In fact, one of the world's most famous psycholinguists, Stephen Pinker, has suggested that one of the main functions of language is that it enables us to engage in conflict resolution. Alternatively, some evolutionary biologists have suggested that language evolved as a means to enter into detailed economic negotiations, while many psychologists believe that it is a fundamental component of thinking, remembering and imagining.

Our ability to decode and interpret a stream of sounds into useful and understandable language is so special that it took computer scientists many years to develop adequate speech-recognition software. Even now, our computers can be fooled. If you have a device that responds to speech, try saying the following: *Wreck a nice beach!*, or even *Reckon iced peach!*. Ironically, you may find that the software interprets both these terms as *Recognize speech* – because language is about so much more than simply turning sounds into words. We use context, memory and old-fashioned commonsense to understand what is being said, not to mention a whole raft of non-verbal signals. The rest of this chapter will examine how language is interpreted and produced in the brain, how damage to the brain can affect language, the importance of non-verbal communication, and the potential benefits of speaking more than one language.

HOW THE BRAIN SPEAKS   145

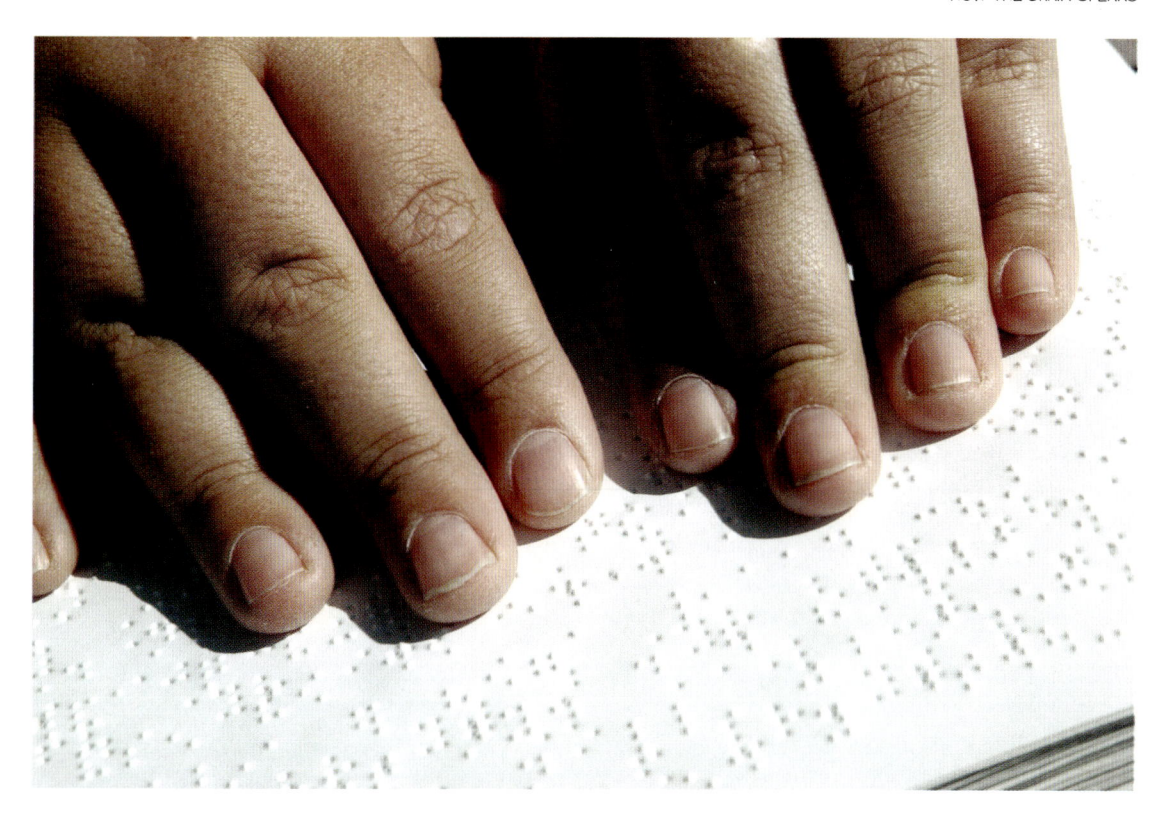

Above: Humans have developed very elaborate systems for preserving the spoken word.

LANGUAGE IN THE BRAIN

Although language was one of the first functions to be localized in the brain, back in the late nineteenth century (*see* page 17), there is still a lot of uncertainty about the finer details of how and where the process occurs. This is partly because language is a complex, multi-faceted skill, which relies on many different cognitive and perceptual processes. Think about the many things your brain has to do in order to interpret a spoken sentence. You will need to: detect the sounds; group them together into meaningful chunks and separate them from all the other peripheral sounds; identify the words (which, as we have seen, will depend upon the context); assign meaning to the words; make sense of how they are sequenced; integrate other important information relating to the way the words are spoken (for example, tone and volume); take into account other non-verbal cues; and then, finally, assign a likely meaning to the overall sentence. And we do all of this extremely competently in a split second! Constructing a sentence is a similarly complex task and yet, once again, we generally do this with great ease, and have the ability to be able to say the same thing in very many different ways.

In Chapter 1, we met "Tan", the patient who progressively lost the ability to speak (*see* page 16). His physician, Paul Broca, published a groundbreaking series of neuropsychological studies on patients with different speech difficulties. These led to the development of some complex models of language function, which still form the underlying basis for the understanding we have today. The clinical term used to describe people who have significant difficulties producing or understanding speech is *aphasia*. Sometimes this impairment is present from birth, but often it occurs as a result of damage to the brain, most commonly through a stroke – a bleed or blood clot in the brain. Every three-and-a-half minutes someone in the UK will suffer from a stroke, meaning a total of approximately 152,000 British people are affected each year. Of those who survive, around half will have their speech affected in some way and at least a third will experience significant clinical impairment in their subsequent ability to communicate. Loss of language can have devastating effects – in terms of overall quality of life, aphasia has been ranked as the third most distressing condition, after cancer and Alzheimer's disease.

THE LESSONS OF APHASIA

But what can neuropsychologists learn from people with aphasia? While most people present with a mixture of symptoms, there are two forms of relatively "pure" aphasia that shed light on the basic processes of understanding and producing speech – these are *Broca's aphasia* and *Wernicke's aphasia* (the latter is named after Karl Wernicke, the German physician, whose work followed on directly from that of Paul Broca). Typically, someone with Broca's aphasia has a reasonable understanding of what is being said to them, but has great difficulty producing words and sentences of their own. They will tend to use very broken speech – often no more than three or fours words at a time – almost as we might if we were trying to communicate in a foreign language that we barely knew. These patients have usually suffered damage to a region of their frontal lobes named *Broca's area*, and it has been postulated that this plays a fundamental role in the production of fluent speech.

Conversely, someone suffering from Wernicke's aphasia can speak quite fluently, but they have a very limited understanding of what is being said to them and, when they do speak, much of what they say sounds like nonsense.

Below: Around half of stroke survivors will have speech and language difficulties.

Fig. 12.1: Broca's & Wernicke's areas

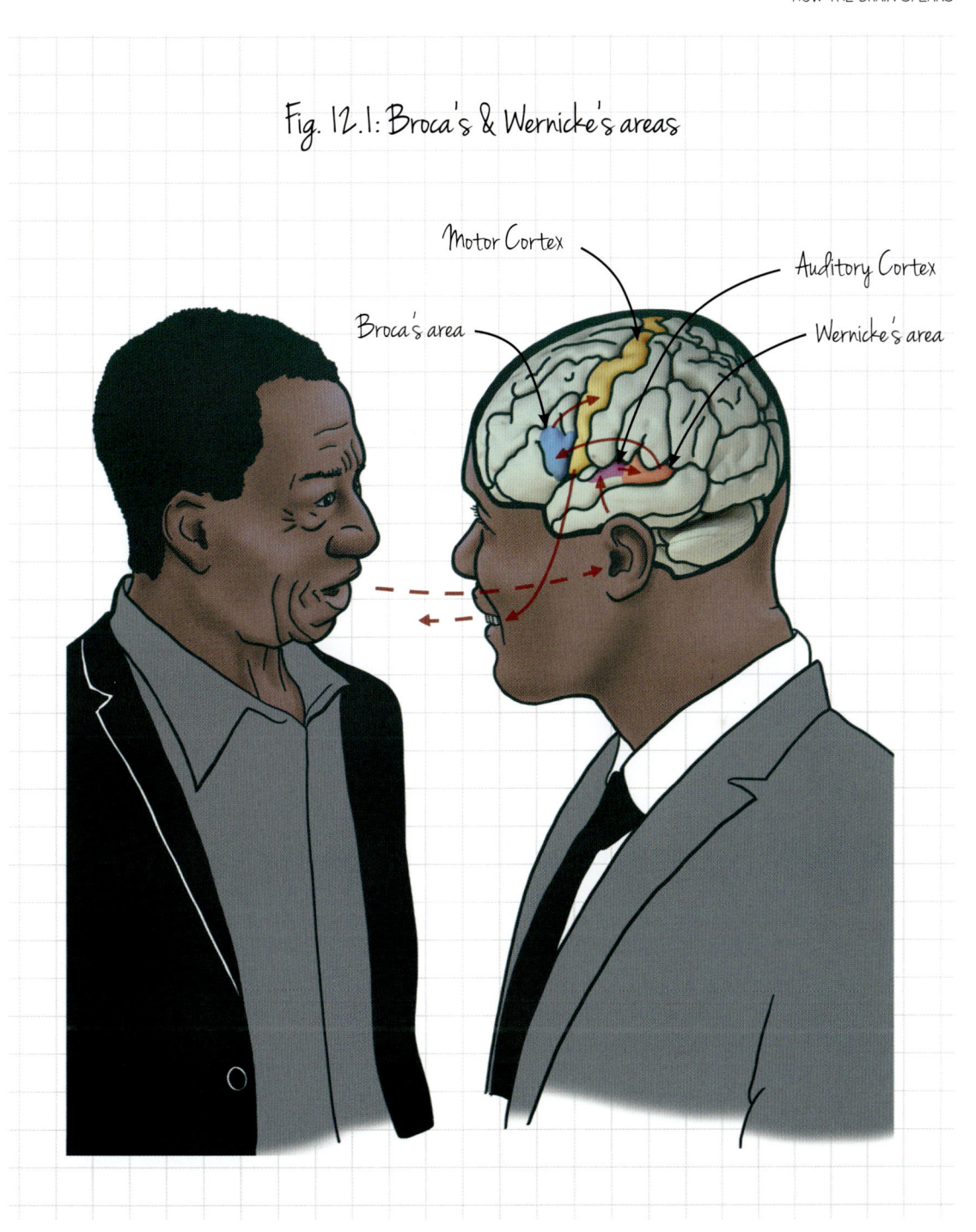

148 HOW THE BRAIN SPEAKS

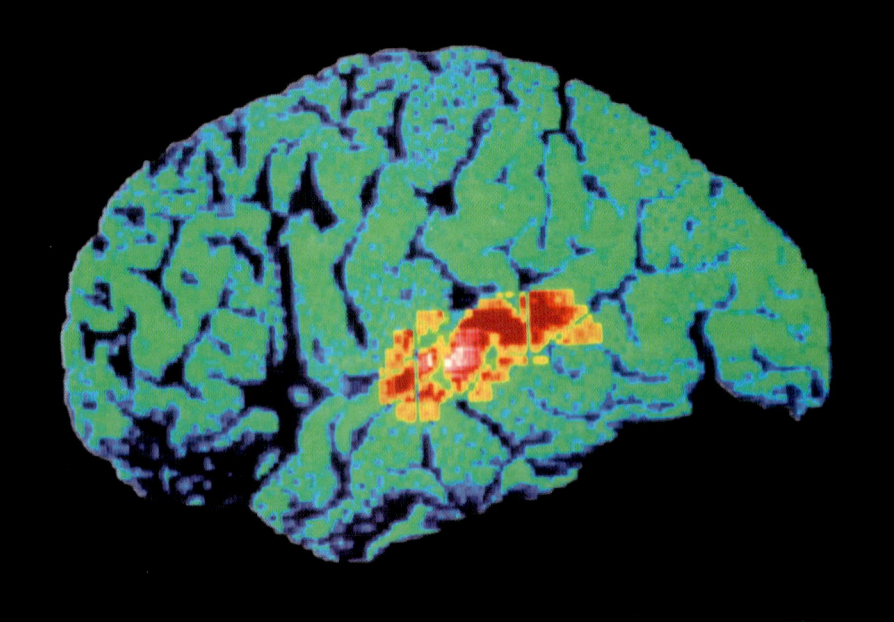

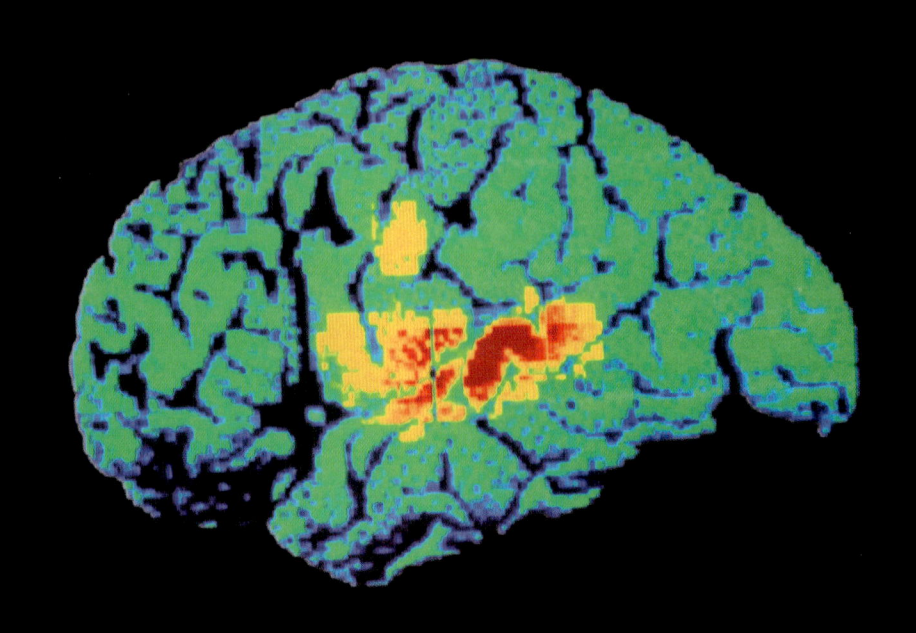

These patients get their words mixed up (for example, *wife* instead of *husband*), they sometimes make up entirely new ones (for example, *gog* instead of *dog*) and, although their sentences are grammatically correct, they often have no coherent meaning. In this case, the primary area of damage is in a region called *Wernicke's area*, which is on the side of the brain where the temporal lobes meet the parietal lobes. What is particularly interesting is that these two language regions can be found on either side of the brain, depending on whether a person is left or right-handed. In approximately 95 per cent of right-handers, Broca's and Wernicke's areas will be on the left side of the brain, but for around 60 per cent of left-handers, the situation is reversed. Contrary to popular belief, this does not mean that language is only found in one side of the brain – simply that one side seems to be more dominant, especially when it comes to producing speech.

While the historical views surrounding the roles of Broca's and Wernicke's areas have continued to be important in our neuropsychological understanding of language, the fuller picture has proved to be somewhat more complicated. For a start, there is no complete agreement on the precise location of these areas and it has become more generally accepted that the traditional Wernicke functions involve a much wider area of the temporal lobes. Thanks to the advent of functional neuroimaging (*see* page 15–16), there has been a surge in the number of investigations over the last twenty years or so, with scientists studying everything from how we recognize words to the mechanisms for understanding and mimicking different regional accents. The current position is that some aspects of language are relatively localized – for example, recognition of single words – but on the whole speech perception and production is a process with many subroutines, and these are distributed across specific widespread networks.

NON-VERBAL COMMUNICATION

In considering which parts of the brain are involved, it is important to bear in mind that good communication relies on far more than just the words we say. Consider the sentence *My computer is broken*. Depending on to whom it is said and the way in which it is said, the same words could mean, *Can you help me?*, *I'm never going to get this work done now!*, or *I'd really like a new laptop, mum*. We make many conscious and unconscious inferences using information from tone of voice, changes in pitch, volume, speed of talking, the use of gestures and the overall context of the utterance. It therefore follows that there is a whole host of additional brain functions that play a role in understanding communication – long-term memory, working memory, emotion, visual perception – to name a few obvious ones. Disruptions to any of these processes can lead to communication difficulties that are quite different to the ones described above – for example those seen in autism (*see* CASE STUDY: DANIEL, page 153), schizophrenia, and Alzheimer's disease.

It is worth examining a couple of these non-verbal elements in a little more detail. If you are someone who normally wears glasses, you may be aware of how much we use our eyes when we listen to people talking. It seems

Opposite: PET scans showing the active areas of the brain when listening to words (top) and when both listening to and repeating words (bottom). A small motor control area is activated when producing speech.

Right: Ventriloquists exploit the fact that speech recognition is partly driven by the mouth movements and gestures we see.

TOP TIPS: SUPPORTING LANGUAGE DEVELOPMENT IN CHILDREN

Early childhood is a period when humans are especially responsive to learning the rules and meaning of speech, and this goes hand-in-hand with the development of other cognitive skills like memory and intelligence. So how can we maximize the opportunities for language acquisition during this critical time?

1. Talk – it is never too early to speak to an infant: give a running commentary of what you are doing; describe and point to things you see; read with them as often as you can (from Day One); and use words to help them identify their feelings – for example, *Oh dear, are you upset / angry?* A recent review showed that *language nutrition* impacts on development – complexity of sentences, quantity and diversity of words, and use of intonation contribute to how a child speaks when older;

2. Eye contact and interaction – children learn best when they are connecting with a speaker and when they get a positive response to their own attempts to communicate;

3. Never criticize – language is best learned passively, through exposure rather than instruction; criticisms tend to make a child self-conscious and more likely to develop a stammer or be shy to speak. An ideal approach is to repeat back what was said but with the correct pronunciation or grammar. For example, if they say, *I writ that story*, you can respond with *Wow, that's great – you wrote that lovely story?*

4. Recognize and properly treat ear infections or hearing difficulties – not surprisingly, any amount of deafness has a direct effect on language development;

5. Engage in musical activities – music is a naturally engaging activity and it is a fantastic way to stimulate and develop crucial auditory skills, such as rhythm and pitch.

Above: Gestures may be as important for the speaker as they are for the listener.

Above: It is never too early to start speaking to an infant, although children learn best with eye contact and interaction.

strange to see someone squinting when they are actually trying to listen to someone speak, but this is because our perception of speech is very much enhanced by being able to see lip movements and facial expressions. In fact, people with hearing impairments often become extremely adept at this skill and are able to depend almost entirely on lip-reading. This multi-sensory effect is so powerful that it can create the unusual *McGurk illusion*, named after British psychologist Henry McGurk. While he was working with his colleague John MacDonald, McGurk discovered a strange phenomenon: if you show people a video of someone saying *ga* but change the sound they hear to *ba*, the overall very convincing impression is that the sound being articulated is *da* – which is not correct on either front! Normally, the visual and auditory signal to the brain are helpfully backing each other up, but in this case they are providing contradictory signals, so the brain has to do its best to make sense of it.

Gestures and other non-verbal clues are also a hugely important part of the human communication system. As we saw earlier, there is some evidence that hand movements are functionally linked with our speech, and this may be why we gesticulate so naturally – and conversely why we find it hard to speak fluently if we are asked to keep our hands still. Many of the gestures we make are quite unconscious, but the real question is: why do we use them? Daniel Gurney, a psychologist from the University of Hertfordshire, has been studying this for a number of years. Gurney acknowledges that when we nod our head, point towards something, wave our hands or fold our arms, it conveys useful additional information to listeners. Certainly there is good evidence that gestures reinforce and embellish the spoken word – they can even be used to induce false memories or to mislead us (a common trick used by magicians). However, Gurney also points out that gestures might have a role in helping us more directly: people find words more easily when their hands are free, compared to when they are restricted, and they also continue to use gestures even when there are no communicative benefits – for example, on the telephone. On balance, it seems that these additional movements may be just as important for the speaker as they are for the listener, providing further illustration of how complex and advanced our spoken language really is.

Below: Speaking two languages may reduce the effects of age on the brain in later life.

DOES SPEAKING MORE THAN ONE LANGUAGE PROTECT THE BRAIN?

Probably the single most impressive thing about human language is the innate capacity we have to learn it – for most typically developing children, learning to speak and understand speech is effortless, automatic and, most would argue, inevitable. The advice generally given to parents is not to correct children when they make mistakes with their speech but rather to answer or repeat the point using correct grammar. This is because children learn language most effectively simply by listening to others and then mimicking them. Interestingly, they will learn faster with parents who make regular eye contact, and one study even showed that the type of buggy an infant is carried in can make a difference; children who faced towards their parent showed better verbal skills, were more interactive and shared more laughter, than those who regularly faced the other way.

By the age of four, the majority of humans are able to communicate their wants, needs and thoughts through spoken language, although interestingly their understanding of speech is nearly always well ahead of their ability to use it. Most parents will have experienced the surprising moment when their offspring suddenly acts on something that has been said to them, despite the fact that that same child is not able to articulate anything much more than *ga* or *da* themselves. What is possibly even more remarkable is that this developmental skill is not limited to just one language. A child who is exposed to two, three or even more languages in the first few years of life will have the potential to become fluent in any or all of those tongues, providing that they continue to be exposed to them. Most research suggests that there is a period within which a child will learn a language effortlessly and with no sign of a "foreign" accent – this extends from birth until somewhere between the ages of five and ten. Beyond this, because the language areas of the brain have matured, individuals are more likely to require explicit training or lessons and will have to apply more conscious effort.

Bilingualism – the ability to speak two languages – has hit the headlines on a number of occasions in recent years. A reasonable body of research has suggested that this skill may reduce the effects of age on the brain and possibly even protect against dementia. At first glance it seems unlikely that simply speaking a second language could have such an impact, but theorists suggest that the continual switching between two different language systems may exercise and develop the prefrontal cortex, an area that ages more quickly than the rest of the brain (*see* Chapter 11). The degree to which bilingualism may offer benefits is still being debated, but there is little question that the simple act of communicating with our fellow humans involves widespread activation of some of the brain's most important networks. Whether we can speak five languages fluently or just one, we are fortunate to have such an extraordinary capacity for sharing, exploring and storing the contents of our mind.

CASE STUDY: DANIEL

Daniel was a beautiful, alert and very active baby. He learned to walk unusually early and could already run across the room by the age of nine months. Sadly, he did not appear to be so quick to speak, and by the age of four he was still only using single meaningless syllables – his favourite expression was *ah, ee, ee, e!* Despite many people telling them not to worry, Daniel's parents instinctively felt something was not right and, after a number of meetings with health professionals and psychologists, he was eventually diagnosed with autism. With support, Daniel learned to speak, but he often used language in an unusual way – for example saying, *He wants to take his shoes off,* instead of *I want to take my shoes off*. He also spoke with little expression and enjoyed repeating words, phrases or questions over and over again. Words tended to have a very fixed meaning for him, which meant that jokes often made no sense to him and expressions such as *He laughed his head off!* could make him very worried.

Daniel is now a young adult and, with many years of excellent support and training, he is able to use language in a way that allows him to communicate and socialize, even though his speech still sounds rather wooden and he can misunderstand things. Despite Daniel's intellectual impairments and difficulties with communication, he is a charming, polite and conscientious man, who is able to live in semi-independent accommodation and work in a hotel three days a week. His achievements are a great testament to his own commitment, his parents' unwavering determination and many years of excellent speech and language therapy.

13. MAKING SENSE OF OUR WORLD

FROM KNOWING WHICH WAY UP WE ARE STANDING, TO IDENTIFYING A COMPATIBLE PARTNER THROUGH A ROMANTIC KISS, OUR ABILITY TO SENSE THE WORLD BOTH AROUND AND INSIDE US IS TRULY IMPRESSIVE.

Our eyes are capable of responding to a single photon, the smallest unit of light, and most of us can recognize more than five thousand different odours. But responsiveness to environment is one of the most basic criteria of life – even the common garden daisy will close its petals when it gets dark – so what, if anything, is special about the highly evolved human sensory system?

Below: Our sensory receptors enable us to experience the pleasure of our favourite music.

SENSING OUR WORLD

Like many animals, we use a range of sensory receptors to keep us safe and well, but these same receptors are also capable of delivering enormous aesthetic pleasure, whether it is in the form of a beautiful landscape, a thrilling piece of music or a delicious gourmet meal. This is all made possible by the highly sophisticated perceptual processes of the human brain. Our senses merely provide information about the world – albeit very rich data – and it is our minds that

MAKING SENSE OF OUR WORLD 155

Above: In this picture you are able to sense a number of different colour dots, but you should be able to perceive the number 74.

transform these into conscious experiences which stir our emotions and guide our decisions.

The distinction between sensation and perception is an important and slightly tricky one. Do we *sense* the smell of leaking gas or do we *perceive* it? From a layman's point of view, the two terms may seem to mean one and the same thing. However, psychologists make a clear distinction between them: *sensation* is defined as the detection of external or internal stimuli; *perception* is the conscious registration and interpretation of that input. The interesting thing about both of the examples at the start of this chapter is that they do not involve any conscious perception. It would be very inefficient if we had to continually waste precious cognitive resources ascertaining whether we were lying down, standing up, facing right or facing left. And yet, even with our eyes closed, we have an extremely accurate awareness of how each of our limbs is positioned in relation to others and the ground. Even in our deepest sleep we sense discomfort or pain and move accordingly. And this is only the tip of the iceberg; our brains can quietly detect and respond to all sorts of internal signals, as well as glucose, oxygen, salt levels, temperature – the list goes on.

PATHWAYS OF SENSATION AND PERCEPTION

Before we return to the intriguing question of how kissing can help us decide on the compatibility of a potential mate, let us look at the basic process that occurs when the brain picks up a change in the environment. The first fundamental requirement is that there must be some kind of stimulus, capable of initiating a response by the nervous system – this may be light, heat, sound, mechanical energy or chemical energy. This is then detected by a receptor – for example, a nerve ending in the skin, or a more complex sense organ, such as the eye. These specialized neural tissues then convert (or *transduce*) the stimulus into a nerve impulse, which travels along sensory neurons into the spinal cord. Sometimes this may trigger a simple reflex response (for example, if you have trodden on a drawing pin or put your hand in very hot water), but most signals continue their pathway up to the brain, to an area in the organ named the thalamus. From here they are redirected to a dedicated processing area in the sensory cortex and are registered as a sensation. Disruption to any part of this route will have a direct impact on experience; so, for example, blindness or visual disturbance may occur if the eye is damaged, the neural pathways are not intact, or there is an area of dead brain tissue in the visual cortex.

In terms of our experience, this is really only the first part of the journey. The information flooding in from our senses has to be processed, integrated, and then compared with our memory bank of previous experiences, so that it can be properly interpreted and appropriately responded to. Our sensory system also has the intriguing capacity to project sensations back to their point of origin. In other words, we believe that we see with our eyes, taste with our tongue and touch with our fingers, and yet all of those actual sensations come from neurons firing inside the head. Occasionally, the neural wiring can get a little mixed up, leading to a condition known as *synaesthesia* – something that the contemporary pop musicians Kanye West and Pharrell Wiliams share with a number of other famous musicians and artists, including Duke Ellington, Vincent van Gogh, David Hockney, Wassily Kandinsky and Marilyn Monroe. It is estimated that around one in every 2,000 people experience this blending of two or more senses, in which sounds, tastes, smells or words may have a specific colour, shape, smell or sound. In fact, one famous synaesthete – the rock legend, Jimi Hendrix – was so taken by the purple colour that he sensed every time he played a particular chord (E7[#9] for those in the know), that it inspired his famous song 'Purple Haze'.

Traditionally, we speak of having five senses – touch, taste, smell, sight and hearing – and indeed this classification is believed to go right back to the time of Aristotle (384 322 BC). However, as we have already seen, our sensory system goes far beyond this, with a vast array of internal receptors that pick up critical information about chemical and temperature balance. There are also specialist *proprioceptors*, which detect the precise tension in all of our muscles and tendons, allowing the brain to calculate the precise coordinates of each of our body parts. This proprioceptive skill is sometimes referred to as the "sixth sense", although others use the same term to describe a special ability to predict the future or know what someone is thinking. In fact, there is nothing mysterious or special about feelings of intuition – they are simply the product of unconscious processing of sensory information. If, for example, you sense that a friend does not seem quite themselves even though they appear to be behaving quite normally, it is most likely because there are subtle changes in their behaviour that you cannot consciously identify but nevertheless provoke an unconscious emotional response. After all, human emotion itself is fundamentally based on detecting internal sensory changes (*see* Chapter 7).

TOUCH

Of Aristotle's five senses, the earliest one to develop in the growing foetus is touch – starting just three weeks after conception. As the Canadian novelist Margaret Atwood so brilliantly put it, *Touch comes before sight, before speech. It is the first language and the last, and it always tells the truth.* We all know how powerful human touch can be, and research backs this up: strangers are more likely to help someone if the request is accompanied by touch, and librarians are seen as being more friendly if they make brief physical contact when a book is returned. These effects have also been shown in waitresses and shop assistants and happen even

MAKING SENSE OF OUR WORLD   157

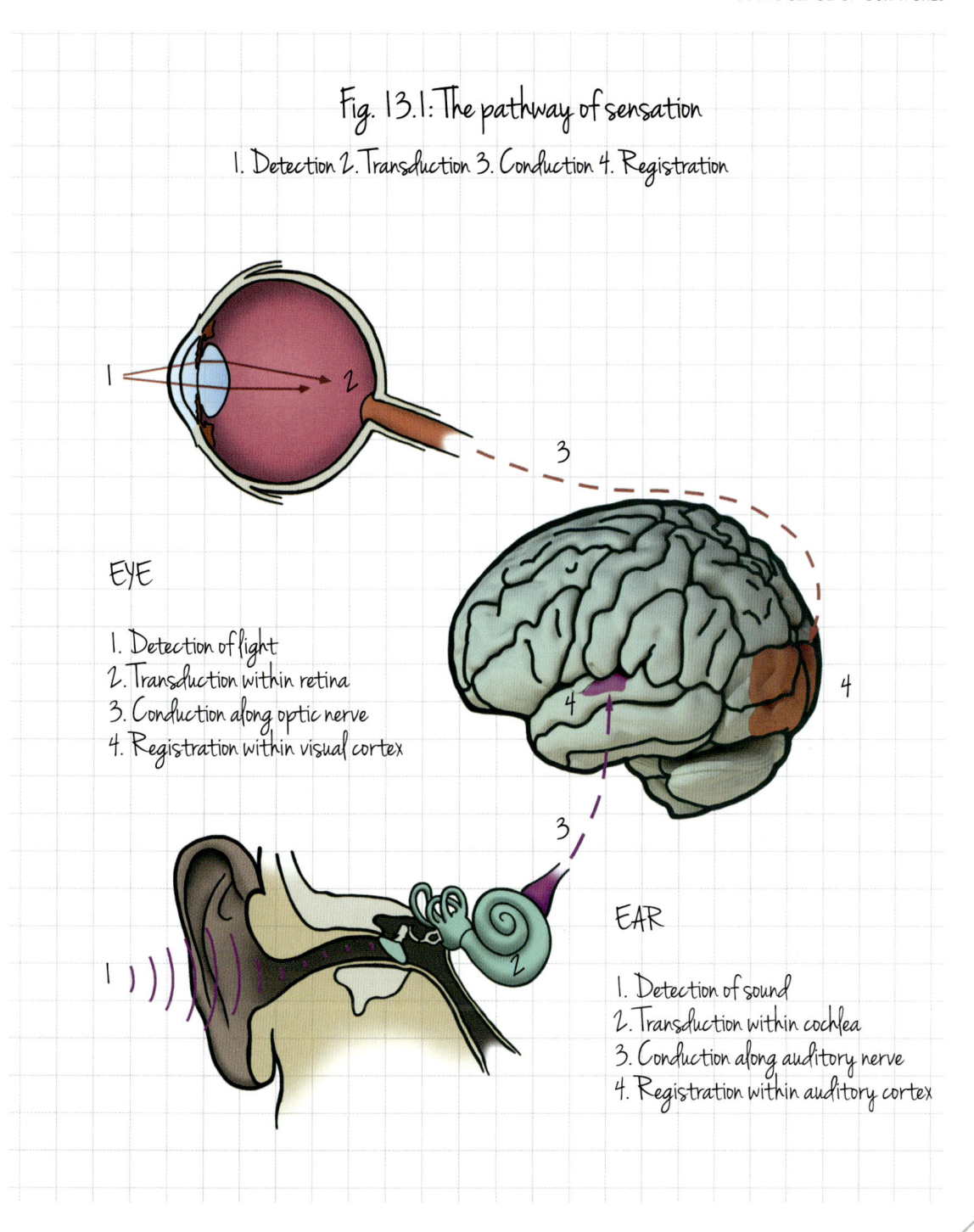

Fig. 13.1: The pathway of sensation
1. Detection 2. Transduction 3. Conduction 4. Registration

EYE
1. Detection of light
2. Transduction within retina
3. Conduction along optic nerve
4. Registration within visual cortex

EAR
1. Detection of sound
2. Transduction within cochlea
3. Conduction along auditory nerve
4. Registration within auditory cortex

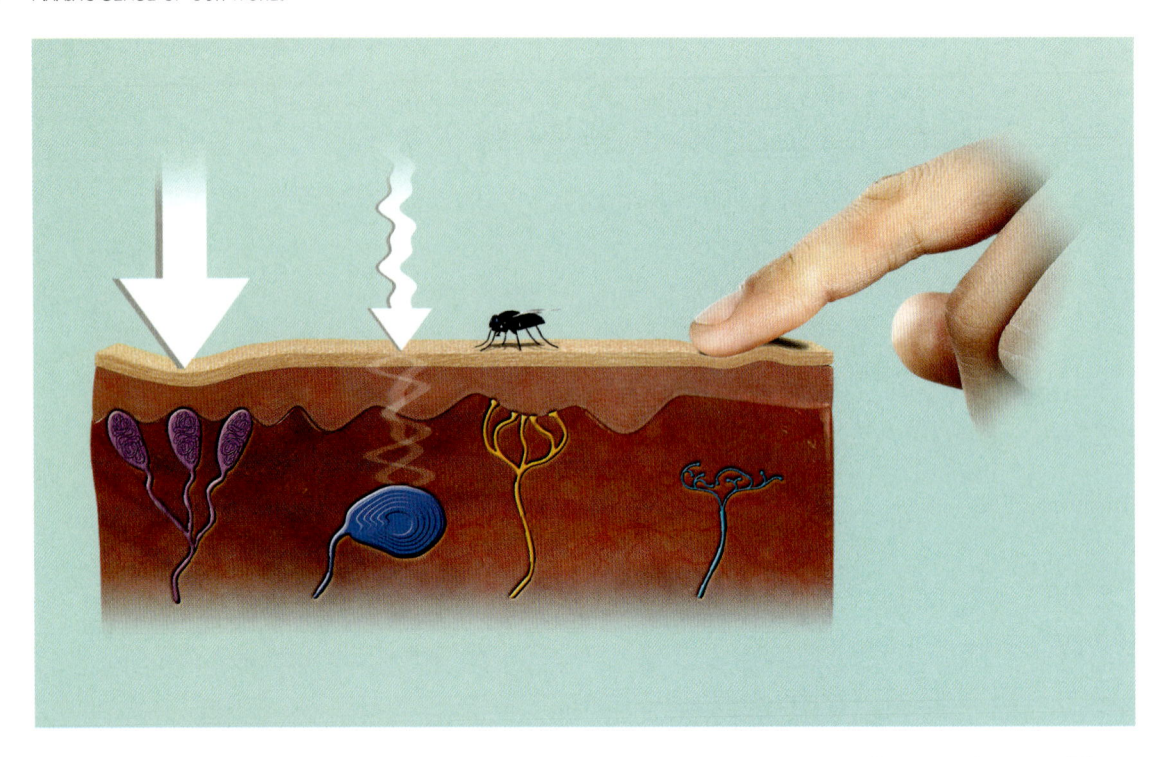

Above: Some of our skin receptors are located near the surface, while others are deeper and only respond to strong pressure.

when people have been unaware of the touch. It can affect team bonding, too – one fascinating study showed that basketball teams which used more "high-fives", hugging and backslaps during the season outperformed those teams that demonstrated less physical contact. Touch can be used to reassure an anxious child, to signal to a friend that we like them and, conversely, to warn someone off if they are getting too close. Psychologists have also shown that touch is an important part of early development, and that babies who are given a lot of physical contact show faster sensory-motor development, better physical growth, a higher level of wellbeing and may even fare better in tests of intellect.

Our sense of touch is remarkably diverse – we are easily able to distinguish between tickling, a gentle handshake and a heavy punch. And some parts of our body are far more sensitive than others – for example, our lips and fingers: think of how easily we can feel the difference between different materials or surfaces, and how deftly a blind person can read the fine dots on a Braille manuscript. This versatility primarily comes down to the wide range of skin receptors that we have and the particular way in which they are organized. Most of these skin receptors have a very simple structure and are easily stimulated by mechanical pressure and vibration, but they vary in their level of sensitivity and placement within the skin layers. Some are located very near the surface or are wrapped around a hair follicle, making them responsive to light touch; others lie much deeper, so they respond only when there is strong pressure. Sensitive areas, such as the fingertips, will have a very high density of receptors, but less important parts of the body like the back or sole of the foot will have far fewer.

Physical stimulation of any of these skin receptors activates the sensory neurons, which then travel up to an area of the brain known as the somatosensory cortex. This is quite a large region, but importantly, every part of the body is directly represented within it, so that it comprises

MAKING SENSE OF OUR WORLD

a complete internal body map. As we know from earlier chapters, the brain is highly adaptable and responsive to its surroundings, so it may come as no surprise to find that areas of the body with more receptors take up far more brain space than those that provide less input (*see* page 160). Given the work of Hubel and Wiesel (*see* pages 126–28), and in light of other research since, it stands to reason that constructing an accurate internal representation of our body relies heavily on the extent of input from touch receptors. This explains why physical contact is particularly important during early development, but also suggests that massage therapy may be a useful approach for people with body image disturbances, such as those suffering from anorexia nervosa. In fact, some theorists have postulated that eating disorders may in part be due to the internal body map not keeping pace with the exceptionally rapid changes that happen to physical form during puberty. For all adults, human touch remains hugely important and beneficial throughout life – it helps us to retain a healthy body image and supports relaxation and bonding through the release of the "cuddle hormone" oxytocin.

Below: Massage has many physical and emotional benefits. It can also play a role in reinforcing or correcting our internal body map.

160 MAKING SENSE OF OUR WORLD

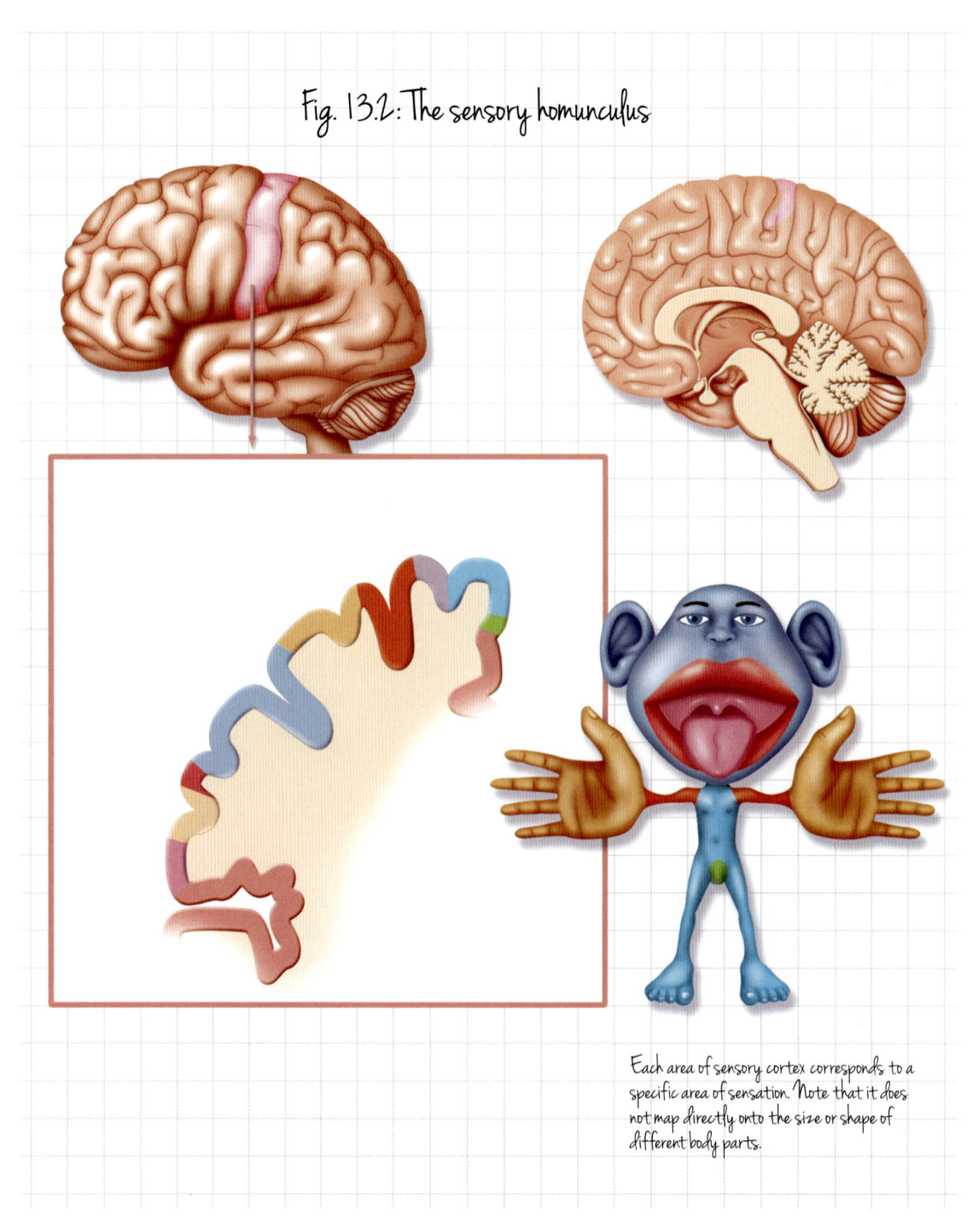

Fig. 13.2: The sensory homunculus

Each area of sensory cortex corresponds to a specific area of sensation. Note that it does not map directly onto the size or shape of different body parts.

TASTE AND SMELL

When it comes to evaluating the value or danger of chemicals in our environment, taste and smell form a formidable and invaluable partnership. Like touch, these senses are fairly well developed by the time a baby is born and they both play a significant role in safety and nutrition, letting us know what is safe to eat and what we should stay away from. Although newborn babies have an innate preference for sweet flavours and a dislike of anything bitter, environment has been shown to play an important part in shaping later food choices. Babies who are given a wider variety of tastes to sample from early on will have a broader palette as they grow older – another illustration of how the brain adapts to its environment. It is interesting to note that this exposure includes breast milk, which automatically carries the taste of what the mother has eaten herself. When it comes to conscious perception of taste, the nose is just as important as the tongue. Anyone who has had a bad cold and a blocked nose will know the frustration of not being able to identify anything they eat. It may seem hard to believe, but experiments have shown that when the sense of smell is removed, it is almost impossible to distinguish between an apple and a raw onion! And it does not stop there, because the taste experience is heavily influenced by many other sensory qualities – for example, temperature and texture, as well as what the food looks like.

Below: Smells can improve mood, trigger memories and warn us of danger.

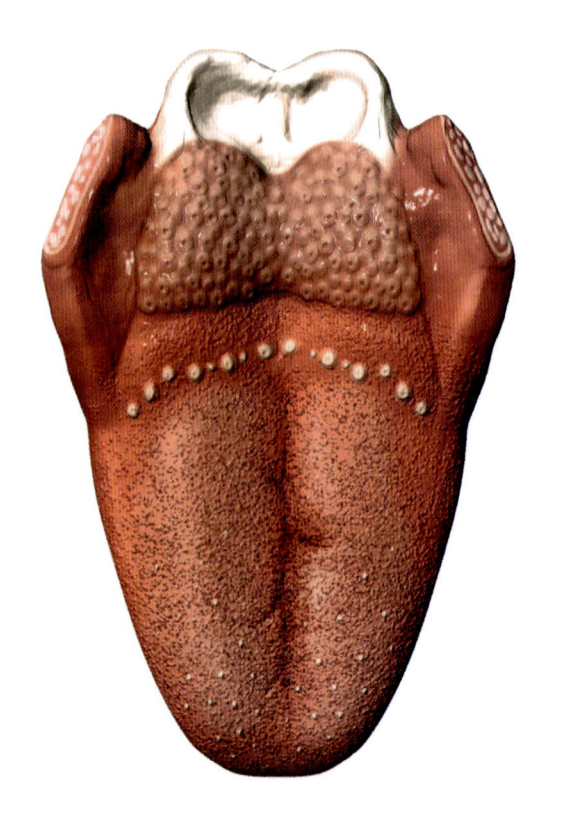

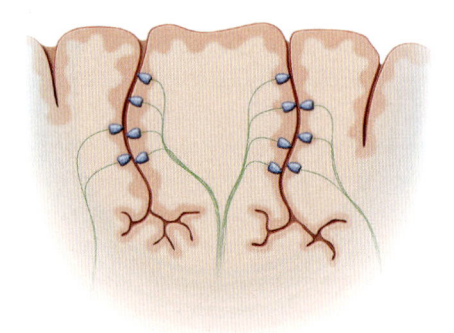

Above: The tongue contains many tastebuds which can detect and respond to a huge variety of different foods.

In addition to influencing appetite and taste, smell plays a key role in human bonding, attachment and sexual attraction. A new baby has an immediate recognition and preference for the scent of its own mother, and older children can reliably distinguish t-shirts that have been worn by themselves or their siblings from those worn by strangers. "T-shirt sniffing" experiments like these have shown that the ability to identify family and partners by smell continues into adulthood. And as we saw at the beginning of the chapter, smell can even help us to identify a suitable partner…

A romantic kiss naturally puts noses in very close proximity to each other, and the Swiss biologist Claus Wedekind has found that women are most attracted to men who carry a contrasting genetic code for immunity. He suggests that mates with complementary DNA are more likely to produce offspring with a stronger immune system, and theorizes that this has an unconscious influence on personal attraction. Interestingly, this effect was reversed in women who were taking the contraceptive pill, suggesting that hormonal influences are also important. Overall, although our sense of smell is nowhere near as acute as that of many other animals, it is still a very important part of social and sexual interaction. And like other senses it shows plasticity – professional whisky blenders and perfumiers have a larger region of their brain dedicated to smell than the rest of us.

MAKING SENSE OF OUR WORLD **163**

SIGHT AND HEARING

Although smell and taste both rely on specialist organs that have evolved to be highly responsive to chemical stimuli, the mechanisms are relatively straightforward compared with those involved in sight and hearing. Our eyes and ears are highly adapted sophisticated structures and the same is true of the corresponding visual and auditory areas in the cortex. From a developmental point of view, hearing matures much sooner than sight, but the visual system catches up quickly and, by the time a baby reaches the end of its first year, it can see almost as well as an adult. Because development of the specialized visual cortex is heavily dependent on input, any early problems with the eye itself – congenital cataracts or a squint – can have long-term effects on vision if they are not corrected early enough. However, when they develop normally, the ears and eyes each work in pairs to provide a detailed three-dimensional landscape of sound and vision.

The human eye has many parallels with a traditional SLR camera – it has a lens that can alter focus, a pupil that can

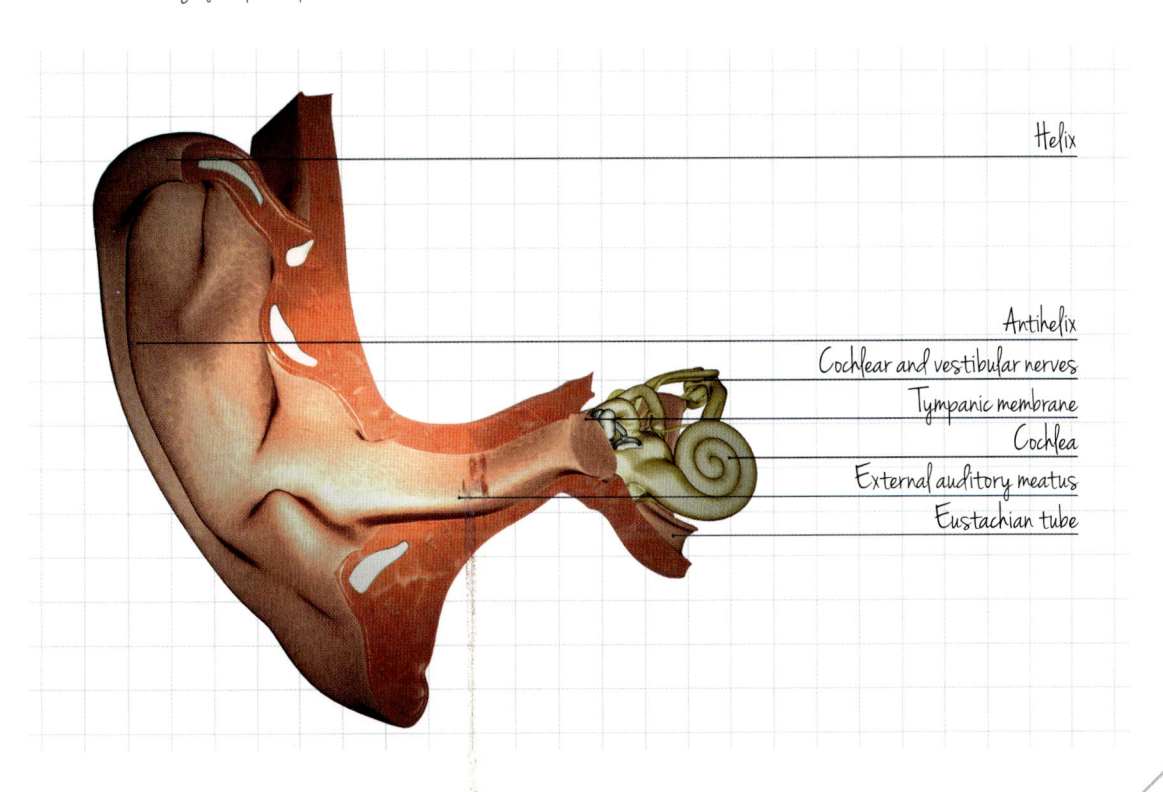

Above: Studies have suggested that couples gather important information about the DNA of a potential mate when they kiss.

Below: The ear is a highly adapted sophisticated structure.

Helix

Antihelix
Cochlear and vestibular nerves
Tympanic membrane
Cochlea
External auditory meatus
Eustachian tube

Above: Mankind's capacity to imagine allows us to create innovative concepts and artworks.

change its aperture to adapt to the light, and a retina that is packed with light-sensitive cells, named *rods* and *cones* (these act a little like the film on the back of the camera). Cone cells have evolved to be specifically responsive to colour and detail, but they become inactive at low levels of light and are unable to fire, which is why everything appears to be some kind of shade of grey at night. These cells are also absent from the peripheral parts of the retina, which means that we can only actually see colour in the part of a scene we are directly focused on. Our visual experience is obviously quite different to this, because we certainly do not feel as though we are looking through a narrow tunnel of colour. This is because the brain is remarkably adept at filling in any gaps in sensory data and continually uses memory to make assumptions and predictions about what we see. It is also why we are so susceptible to illusions. For example, you may have noticed that whenever the moon is near the horizon it

the fact that it forms exactly the same size image on the retina wherever it is. There are various explanations for this phenomenon, but they all relate to us making assumptions based on previous knowledge – for example, believing the moon to be closer to us when it is on the horizon and therefore judging it to be bigger, especially in relation to the trees or buildings in close proximity.

What has hopefully become clear throughout this chapter is that our sensory receptors – however simple or complex they may be – carry out the essential task of collecting and feeding in a rich set of data to our brain about our internal and external environment. But what probably makes human beings stand apart from the rest of the animal kingdom is what we do with that data. We use memory, experience and previous experiences to form a record of the past. Many would argue that the capacity to imagine is unique to humans; we can easily conjure up the delights of a sticky chocolate cake or imagine a loved one's face. Neuroimaging studies reveal that just thinking about a particular smell or sound is enough to trigger activity in the relevant parts of the cortex. Interestingly, hallucinations also activate the same areas that a real perception would (*see* Chapter 14 for details), demonstrating the power of the human mind to create experience. It is true that we would be nothing without our senses, but we would certainly not be human without the capacity to process intellectually, store and act on sensory input. As the famous eighteenth-century German philosopher Immanuel Kant said, *All our knowledge begins with the senses, proceeds then to the understanding, and ends with reason. There is nothing higher than reason.*

Right: Whenever the moon is on the horizon it appears much bigger than when it is high in the sky.

CASE STUDY: HELEN KELLER

Helen Keller was born on 27 June 1880, in Tuscumbia, Alabama, USA. In January, 1882, she became very ill with what her doctor described as "brain fever", but which we now believe may have been meningitis or scarlet fever. As the infection subsided, her mother noticed that Helen was no longer responding to sound and seemed oblivious when a hand was waved over her face. Sadly, the illness had plunged Helen into a lifetime of darkness and silence. By the age of seven, Helen had unsurprisingly become a difficult child – frustrated, angry and unruly. Although her parents were advised to put her in an institution, they believed a better life was possible for their daughter and they searched for a solution.

Help came via none other than Alexander Graham Bell, the inventor of the telephone, who was at that time working with deaf children. He put them in touch with the Perkins Institution for the Blind in Boston, Massachusetts, who recommended a teacher, Anne Sullivan, herself severely visually impaired. Despite a difficult start, Anne made a breakthrough in connecting with Helen and was able to start teaching her new ways to communicate by spelling words out on her hand. With Anne's help, Helen set out on a new journey, which saw her graduate in 1904 – becoming the first blind person to earn a Bachelor of Arts degree – and ultimately led to her becoming a leading humanitarian and political activist. Helen Keller died shortly before her 88th birthday, leaving behind an inspiring story of how courage and determination can allow a person to understand their world using only the sense of touch.

14. ALTERED STATES OF CONSCIOUSNESS

TAKE A MOMENT TO CONSIDER YOUR CURRENT MENTAL STATE: IT STANDS TO REASON THAT YOU ARE CONSCIOUS WHILE YOU READ THESE WORDS, BUT WHAT DOES THAT EXPERIENCE FEEL LIKE?

As the famous American psychologist William James once said, consciousness exists in many forms and our normal, waking, rational consciousness is just one of these. So, what does it mean to experience an *altered state of consciousness*? People typically assume that this question relates to an encounter with psychoactive drugs, particularly those that are hallucinogenic. However, drug-taking provokes just one very specific experience of consciousness: if you have ever had a high fever; been excessively tired or jet-lagged; drunk a few glasses of wine; or taken part in a meditation or hypnosis session – then you will almost certainly know what it feels like to take a slight detour from the normal waking state. Any of these experiences would leave you feeling maybe a little light-headed and detached, with your senses slightly sharpened or dulled, or possibly with a distorted sense of time. Individual experiences vary hugely but, put simply, an altered state of consciousness encompasses any experience that deviates from your "normal" state of mind.

Below: According to German psychologist Dieter Vaitl, everyday consciousness is just the top of the iceberg.

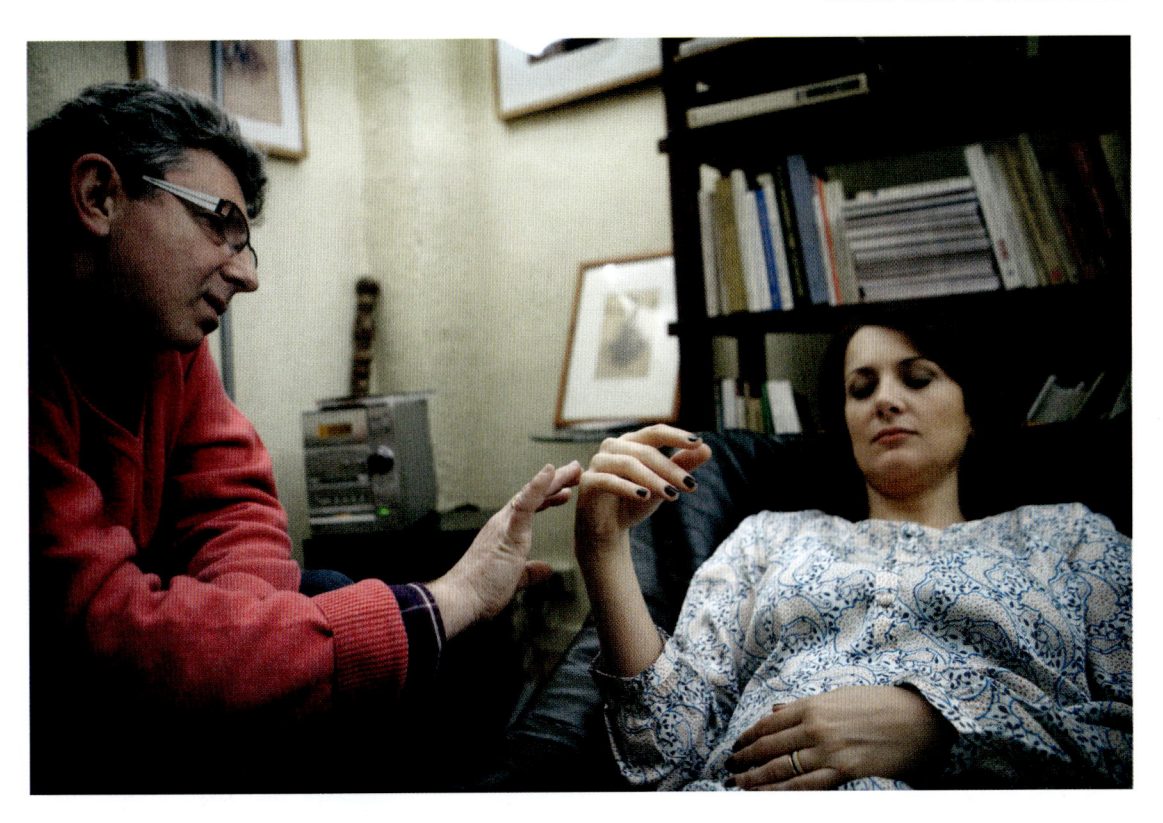

Above: Does hypnosis create an altered state of consciousness? This is something that is hotly debated among scientists and philosophers.

DEFINITIONS AND ASSESSMENT

The notion of an *altered state of consciousness* (ASC) was first brought to public attention in 1969 by the American psychologist Charles Tart. He pointed out that in some cultures, entering into a trance state and "being possessed by a god" is considered quite normal – in fact, those individuals who cannot do this are sometimes seen to be psychologically lacking. Tart also noted that in the Indian language Sanskrit, there are around twenty words to describe different shades of consciousness or mental states, and he suggests that western cultures are generally quite naïve and unsophisticated when it comes to recognizing the breadth of conscious human experience. And yet in the west we do have words for some of the phenomena that occur, for example: *hallucinations, dissociation, lucid dreaming* and *hypnagogia* – the strange perceptual sensations that some people have just before falling asleep. From a scientific point of view, defining an *altered state* is quite tricky, mainly because despite centuries of philosophizing, there is still no consensus on the definition of consciousness itself.

From a medical perspective, consciousness is defined and measured through observations of behaviour and brain activity – is the individual awake and aware of their surroundings, are they responsive, do they have normal reflexes? EEG recordings and other types of neuroimaging can also give a good idea of activity in different regions (*see* Chapter 1). So, maybe we could use a similar approach to assessing whether someone is experiencing an altered state of consciousness? In some cases there are obvious behavioural clues – for example, the slurred speech and changes in demeanour that define drunkenness, or the tiny pupils and glassy eyes that occur in someone who has taken heroin. But what about a person who is meditating

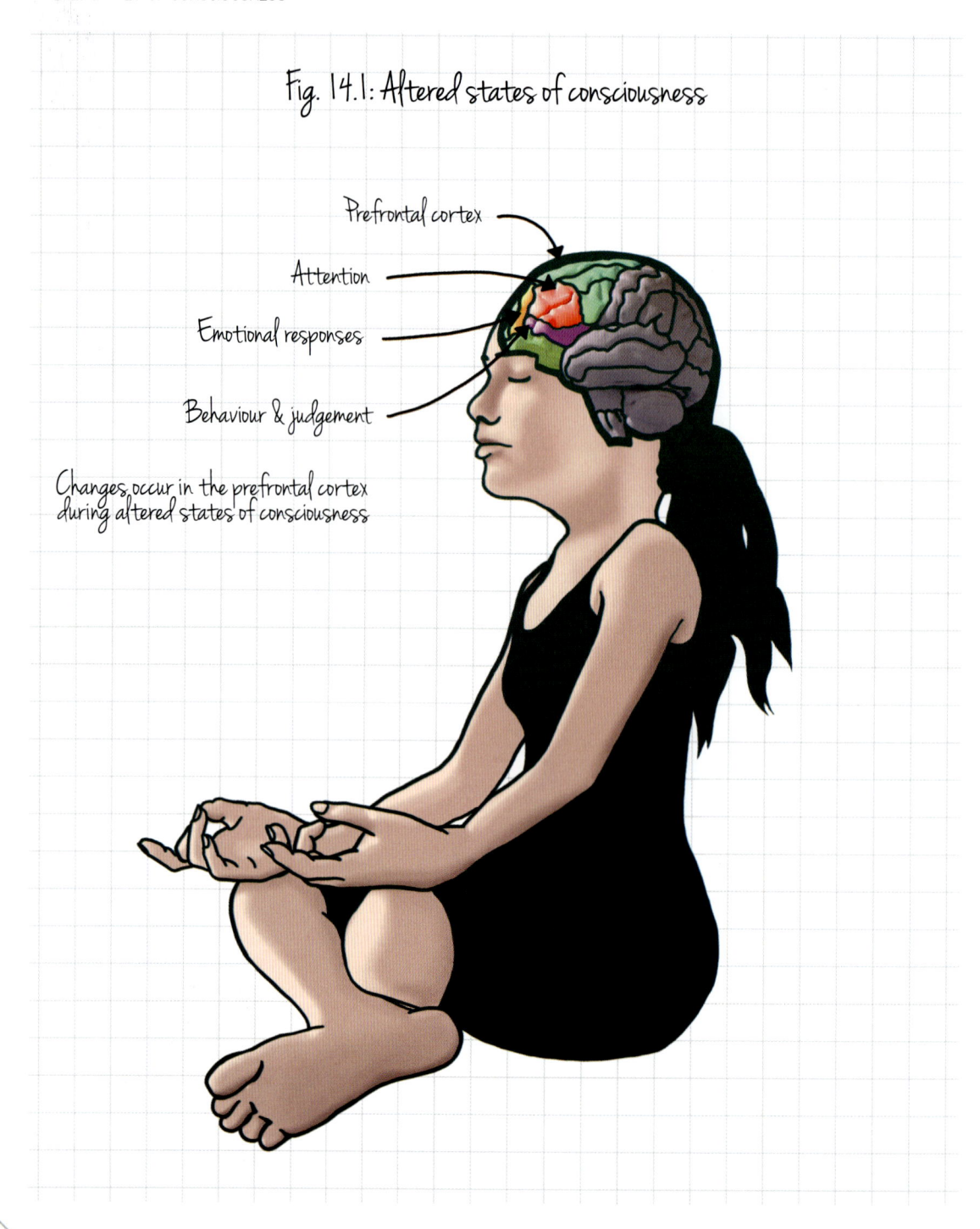

168 ALTERED STATES OF CONSCIOUSNESS

Fig. 14.1: Altered states of consciousness

- Prefrontal cortex
- Attention
- Emotional responses
- Behaviour & judgement

Changes occur in the prefrontal cortex during altered states of consciousness or on the edge of sleep? Because there are so many different experiences that characterize an altered state, there can be no definitive objective assessment, but we do know that there are often changes in EEG activity and blood flow, as well as fluctuations in the levels of certain neurotransmitters.

A more popular strategy is to take a subjective view, so that an altered state is described in terms of the way an individual *feels*. This makes more sense, given that Charles Tart's original definition requires someone to personally experience a radical change in mental state. This may include an alteration in attention, inner speech, time perception, suggestibility, imagery, thinking, body image, sense of identity, or indeed any other aspect of conscious experience. But how can we distill all of these into a viable way of quantifying an altered state? British psychologist Susan Blakemore has spent much of her career exploring the nature of consciousness, from both an academic and phenomenological perspective. She suggests that a useful approach is to focus on changes in three domains – *arousal*, *attention*, and *memory* – so let us look briefly at each of these.

AROUSAL, ATTENTION AND MEMORY

Consider a time when you might have experienced an altered state yourself – maybe something relatively common, like the borderline between wakefulness and sleep, when you are still aware of your surroundings but feel yourself slipping into "nonsense land"; or, maybe a time when you have gone without food for a long time and the world around you feels a little fuzzy around the edges. Let us reflect on Blakemore's first variable, overall *arousal*. This essentially describes how awake, alert and stimulated you feel. In some cases arousal might be very low – for example, in a state of deep relaxation or if a sedative has been taken. Alternatively, there may be times when arousal is disproportionately high, when a stimulant drug has been consumed or in the middle of a bout of extreme exercise. Either way, a change in arousal has the potential to influence our overall mental state in quite dramatic ways, leading to a knock-on effect on all of our thoughts, feelings and perceptions.

What about *attention*? According to Blakemore, this can change in two ways during an altered state. Firstly, the spread of our attention can change so that it becomes very broad and unfocused, flitting from one thing to another very quickly. Or conversely it might become incredibly narrow, bringing one sound or thought into sharp, intense, unwavering focus. Attentional direction can also be influenced. In a typical alert mental state, attention will continually and appropriately switch between internal thoughts and a perception of the external world. However, in an altered state we can become excessively inward or outward facing. An internal focus leads to a feeling of distance from all that is going on around you and a much greater awareness on inner thoughts and sensations. Alternatively, attention can become entirely directed toward the outside environment – for example, focusing on a flower or a pattern. In fact, this phenomenon is often exploited in techniques that actively induce an altered state – for example, the use of an object or picture as the focus of meditation practice.

Memory is inextricably bound up with our conscious experience, not least because our window of the here-and-now is almost entirely determined by the size of our short-term memory. This is defined by how much we can hold in our mind at any one time, rather like the RAM of a computer or the screen on a calculator. Anything that temporarily shortens or extends the span of our short-term memory will change the size of our window of consciousness, which in turn distorts our perception of time passing. The impact of memory goes beyond this, though: you may take for granted how much time you spend reflecting on past events or imagining things that might happen, but research shows us that this kind of mind travel is a very dominant feature of our conscious experience. It therefore follows that anything that limits or changes the way we access and construct memories will influence these processes. This is quite nicely illustrated when we look at different types of dreaming. Daydreams, lucid dreams and nocturnal dreams all use the building blocks of memory to create imagined experiences, but the nature of these dreams varies quite dramatically in different states of consciousness. Blakemore points out that combined changes to memory and attention can force people to be very focused on the present moment, which some may find quite liberating but which might also have negative consequences (*see* CASE STUDY: CLAIRE, page 177).

SPONTANEOUS AND PHYSICALLY INDUCED ASCS

So far, we have seen that it is possible to quantify an altered state of consciousness using a physiological assessment or subjective report. However, Dieter Vaitl, a German psychologist from the University of Giessen, suggests that it is helpful to examine these experiences in relation to how they are induced: spontaneously, physically, psychologically, through disease or brain damage, or as a result of psychoactive drugs. Each of these has its own distinct influence on the brain, which can explain the breadth of the psychological and emotional changes that they evoke. Take sleep, the most common form of altered state: brain activity follows a predictable pattern, cycling between periods of slow to high amplitude waves when we are in deep sleep, to higher frequency, low amplitude waves during periods of dreaming. Levels of cognitive arousal also follow a cyclical pattern – very low in deep sleep, slightly higher during light sleep, and potentially much higher during dreaming.

Dreams are a strange state of consciousness – we are aware of very vivid thoughts and feelings and yet we are asleep! Whenever we dream, the brain stem blocks all outward neural signals to the muscles – a mechanism that paralyzes the body so that we cannot physically respond to whatever bizarre things we might be thinking, feeling or seeing. Therefore, despite our brains being active while we dream, most of our body tends to be relaxed. The one exception is the eyes, which make the random rapid eye movements (REM) that characterize the dream stages of sleep. Attention, unlike in our waking state, is firmly focused inwards. Memory does strange things, too, with bits and pieces of past experience and knowledge thrown together into fantastic, implausible concoctions that can often seem sensible at the time. *"Dreams, if they're any good, are always a little bit crazy"*, said Ray Charles, the legendary American pioneer of soul music.

HYPNAGOGIA AND HYPNOPOMPIA

The passage into and out of sleep has its own strange characteristics. Around 40 per cent of adults have experienced *hypnagogia* – unusual hallucinatory or dream-like perceptions that can occur around the time of sleep onset (or *hypnopompia*, for when it happens during waking). People have described a whole range of peculiar sensations and perceptions, including feelings of flying and floating, or seeing repeating patterns that seem to ebb and flow. This might even include the relatively common *Tetris effect* – the sensation of continuing to do a repetitive activity you have been doing during waking hours, such as driving on a motorway, digging the garden, or playing a computer game like Tetris! Some writers have suggested that hypnagogia has strong similarities to the effects of

Opposite: Daydreaming is a common altered state of consciousness.

Below: Astronauts can experience altered states of consciousness during training and flights due to changes in brain oxygen.

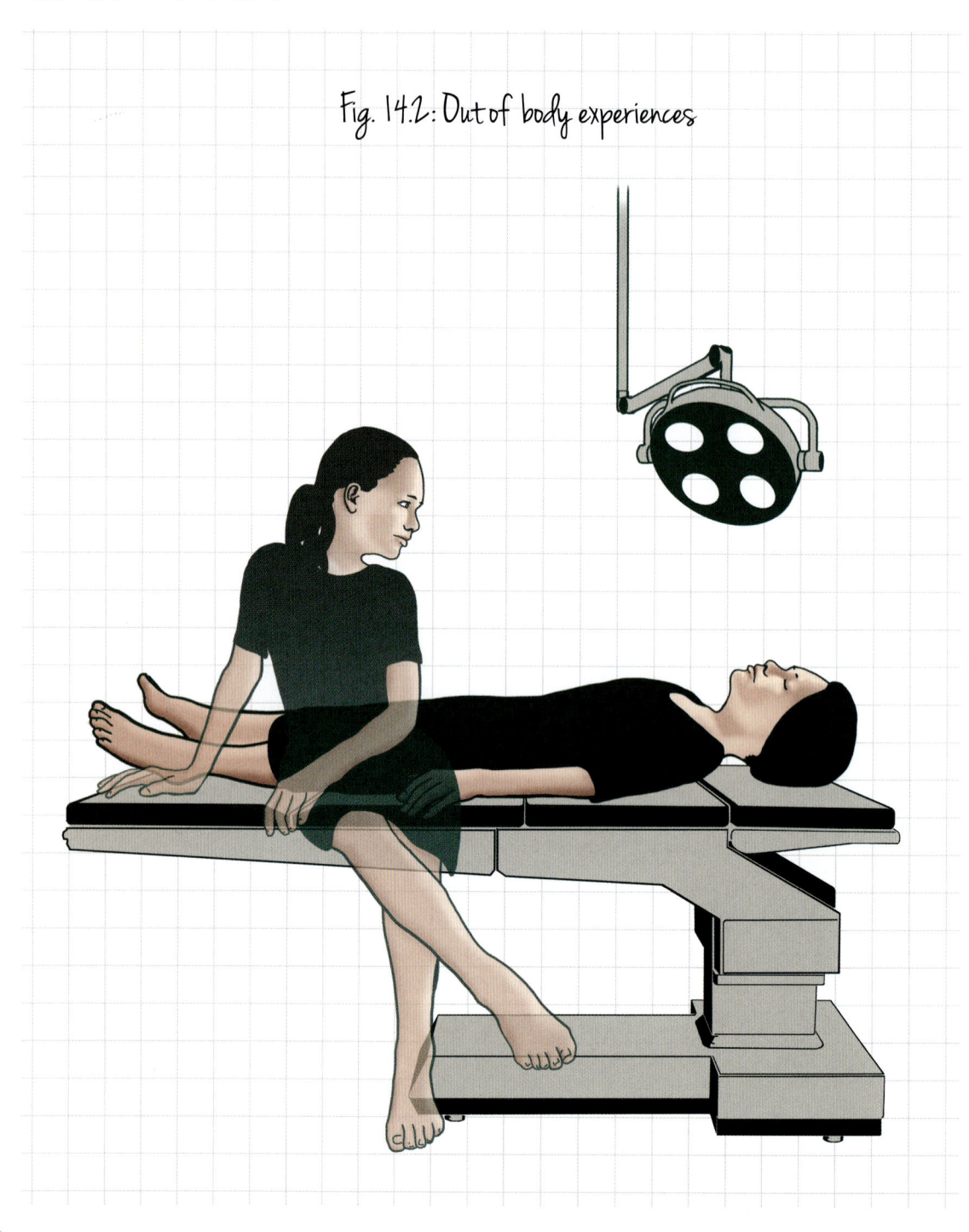

172 ALTERED STATES OF CONSCIOUSNESS

Fig. 14.2: Out of body experiences

LSD or other psychedelic drugs, and acknowledge that this gateway to sleep can range from pleasant to uncomfortable, and fascinating to utterly terrifying. Indeed, for some children this phenomenon can be a real cause for them to dread their bed, and yet there are adults who put huge effort into developing ways of extending and indulging in this hypnagogic period. From a brain point of view, EEG readings show very specific activity during these moments – flattened alpha waves alongside an increasing of slow theta waves (*see* page 18). The most popular explanation is that this very specific altered state occurs when people enter REM sleep so quickly that they effectively start to dream while they are still conscious. Because the dream essentially overlays observed reality, it appears to be real, thus creating an hallucination.

Researchers who carry out any kind of EEG experiment will tell you that even during a normal waking day it is quite common for the brain to drift into brief moments of sleep-like activity. These scientists become quite adept at spotting the signs that someone is beginning to lose focus or becoming bored with their experiment and drifting off. This illustrates that it is not uncommon to experience transient episodes of the crossover between wakefulness and sleep – sometimes called a *near sleep experience*. Of course, these symptoms are more frequent in people who are excessively tired, and this can create its own brand of altered experience. Whether it has been caused by an overzealous social life, a baby who will not sleep, or an urgent deadline, most of us know what it is like to feel sleep-deprived. People describe a narrowing of attention, impairments in memory, difficulties concentrating, distorted time perception, light-headedness, occasional elation and, sometimes, hallucinations. There are also measureable changes in physiology, which include slower brain waves and an increased variability in pupil size.

Other spontaneously induced altered states range from everyday daydreaming (*see* CASE STUDY: CLAIRE, page 173) through to more unusual phenomena, such as out-of-body and near-death experiences. The latter are particularly difficult to investigate, but are also utterly fascinating. They occur in people who have come close to death and have been revived – in fact, EEG recordings often show a complete absence of any activity in the cortex of the brain for a short period. Narrative accounts include feelings of peacefulness and wellbeing, a sense of being outside or above the body, dark tunnels with a brilliant light at the end, mystical feelings of love and union, sounds of music or distant voices, a slowing of time and a speeding up of thoughts. There are several hypotheses for why these occur, most popularly that there is a dramatic loss of oxygen in the brain but also a depletion of neurotransmitter reserves and enhanced release of endorphins (neurotransmitters similar to heroin, which are usually released to ease pain).

In reality, any extreme physiological challenge can cause marked changes to our conscious state. Changes in breathing patterns provide an excellent common example of this – from the effects that can be brought on by yogic or meditative breathing, to the strange feelings of de-realization that can occur during a panic attack. Even a few minutes of hearty singing can provoke a euphoric or slightly light-headed moment. What these activities have in common is that they alter the balance of oxygen and carbon dioxide in the blood, which has a direct impact on brain function. Extreme changes in altitude can also cause some strange psychological sensations – feelings of bodily distortions, difficulties concentrating, as well as visual and auditory hallucinations. In fact, an altered state can be brought on by anything that is associated with sudden metabolic changes or disruption to the flow of oxygen and glucose to the brain. This includes extreme starvation or pathological dieting (although it is not uncommon to experience some degree of light-headedness simply by missing a meal). Interestingly, there is even evidence of specific EEG patterns and a brief partial loss of consciousness during orgasm.

DELIBERATELY INDUCED ASCS

Sometimes people make a more active decision to seek out an altered state of consciousness. At the height of their fame in 1967, the Beatles attended a lecture by a Transcendental Meditation teacher, Maharishi Mahesh Yogi. So fascinated were they that they learned how to meditate and soon afterwards took themselves off for a month-long retreat in India. Nowadays, there is a whole raft of different meditation

Above: Short periods of sensory deprivation can have positive effects on some people.

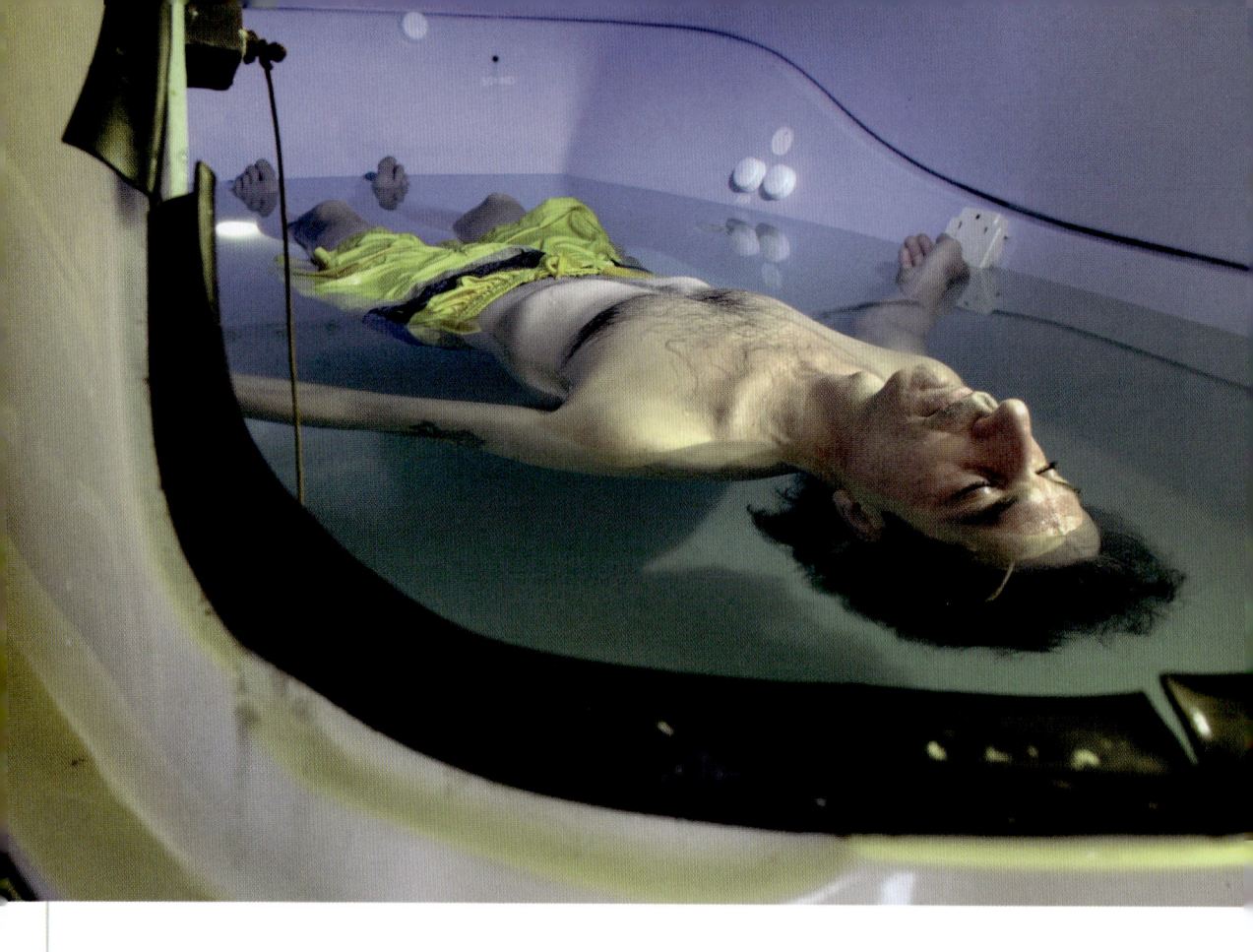

practices that are regularly taught and practised in the western world, not to mention yoga and other forms of active relaxation. All of these are capable of influencing Blakemore's three pillars of consciousness – attention, arousal and memory. Another simple but surprisingly powerful technique is sensory deprivation, the deliberate removal of all forms of sensory input. By denying the brain any external input, this practice forcibly directs all attention to our inner thoughts. In the short-term, this can have quite positive effects on some people, leading to improved memory and creativity, lower levels of adrenaline and cortisol, and increases in endorphins. However, over longer periods of time it can cause a lot of mental distress, inducing anxiety, strange thoughts and hallucinations. This can be so unpleasant that it is an effective form of torture, as many regimes have discovered.

By all accounts, transcendental meditation was not the Beatles' first foray into deliberately induced altered states. Between them, John, Paul, Ringo and George have admitted taking amphetamines, cannabis and LSD – all psychoactive drugs that have a profound effect on mental functioning. John Lennon is quoted as saying, *We were smoking marijuana for breakfast. We were well into marijuana and nobody could communicate with us, because we were just glazed eyes, giggling all the time.* As we saw in Chapter 3, mankind has been experimenting with mind-altering substances since the dawn of time, and across pretty much every culture. Despite the physical and psychological risks that some of these drugs carry (with alcohol arguably at the top of the list), they continue to be grown, produced, sold and used in large quantities across the globe. We have seen in Chapter 3 that psychoactive drugs stimulate the inherent reward systems in the brain, but why do they induce an altered state?

Stimulants – which include caffeine, nicotine, cocaine and amphetamines – have a very direct effect on overall arousal. By increasing levels of excitatory neurotransmitters, they essentially turn the volume up on all of the senses and increase the speed at which thoughts are processed, which in turn can influence time perception. Stimulants often increase attentional focus, as well. Conversely, drugs such as alcohol and benzodiazepines act as a depressant on the central nervous system by enhancing the effects of inhibitory neurotransmitters. They slow down all neural communication, thus lowering overall mental arousal, which causes feelings of relaxation and drowsiness, as well as a reduced control over thoughts and actions. Such is the power of these depressants that excessive use can lead to a state of complete unconsciousness (coma) and even potentially death.

Over the years, hallucinogens have probably proved to be the most popular drug-induced approach for investigating altered states of consciousness, appealing to scientists, artists and authors alike. In 1938, a Swiss scientist named Albert Hofmann manufactured lysergic acid diethylamide, now commonly known as LSD. This was part of a bigger medical project to develop drugs that could stimulate breathing and circulation. The drug was put aside for five years, but when Hofmann started work on it again, he accidentally ingested a small amount and quickly found that it gave him a short-lived mind-altering and pleasant experience. A few days later he intentionally tried a small dose again and systematically noted all of the effects. He is reputed to have ridden home on his bicycle as the drug effects started to kick in, hence the vernacular expression *acid trip*.

Below: Psychedelic drugs have a profound effect on mental functioning and can cause hallucinations.

Another hallucinogen, mescaline, was famously sampled by the author Aldous Huxley and written about in his book *The Doors of Perception*. He gives a vivid and captivating description of his experience – his changed perceptions of everyday objects, a new fascination with things that had seemed ordinary a few hours ago, feelings of insight and an unusually altered sense of time. Other drugs have similar effects – for example, psilocybin and ketamine – so the hallucinogenic or *psychedelic* class of drugs now includes any substance that disrupts sensory perception, causes hallucinations or has psychedelic effects. The literal translation of *psychedelic* is "mind-revealing", so the term is generally used to describe experiences in which people have sudden moments of insight and see things about themselves or the world in a new way. Although this term is used interchangeably with *hallucinogen*, hallucinations have a different and very specific definition: they refer to any perception that feels as though it is really happening but occurs in the absence of an appropriate stimulus. Although everyone has a different experience, most hallucinations follow relatively predictable forms – pulsating lights, geometric shapes, intense colours, rotating images, and large domed rooms. It is thought that these might all reflect the architecture of the visual system.

So, while hallucinations are commonly thought of as a symptom of psychosis, they can also occur with drug use, after sleep deprivation, during sensory deprivation, or even when people are just about to fall asleep. In other words, hallucinations and changed sensations are a relatively common element of any altered state of consciousness. Sometimes it is not clear why they happen, but in the case of psychoactive drugs, it seems that there are direct effects on the neurotransmitters – particularly serotonin – so that the sensory pathways are hijacked and altered. There are also suggestions that hallucinogenic drugs may affect the sleep centres in the brain stem in such a way that the boundaries between dream-like experiences and reality become confused, much as they do in hypnagogia. There has recently been a new surge in research around the psychological and neurological effects of this class of drug. As well as providing important insight into the nature of consciousness, there are some increasingly strong arguments for their controlled use in specific therapeutic settings. While this will no doubt remain a highly controversial topic, it seems that the search for the "perfect" altered state of consciousness will continue, both in and outside the science laboratory. And even for those who do not actively seek these more intense experiences, there are many other ways in which consciousness can be altered. So, next time it happens, focus on exactly how it feels and why it is different, and then reflect on what that tells you about the essence of that all-important human quality, consciousness.

Above: Aldous Huxley famously described his experience of hallucinogenic drugs in his book The Doors of Perception.

Opposite: "Magic mushrooms" contain the chemical psilocybin which can evoke hallucinations.

CASE STUDY: CLAIRE

Eight years ago, Claire developed a severe case of amnesia, following infection from viral encephalitis, which destroyed a large portion of her right hippocampus and surrounding areas of the brain. Claire has retained some knowledge about her life, but most of her memories have been wiped away and her ability to make new ones is very impaired. An event that happened a few hours ago will be as vague and nebulous for her as a distant memory is for the rest of us, and the majority of her memories drift completely away within a day or two. Like many amnesiacs, Claire has found that those moments in which she has nothing specific to focus on can be quite uncomfortable and often distressing. For most of us, these are the times when we find ourselves daydreaming – pondering a conversation we had with a friend, reflecting on a childhood memory, imagining a holiday, or simply wondering what we might have for dinner. This "time travelling" is a fundamental and very important aspect of our conscious experience, but for Claire it is missing. Her limited access to memory means that she cannot idly flit around the past and future like the rest of us. Instead, she is stranded in the here-and-now – arguably in a permanently altered state of consciousness – which leaves her with a distorted sense of time passing and an enduring sense of uncertainty. Claire is an intelligent and positive woman who has found her own ways of dealing with these challenges, but we should never underestimate the freedom bestowed upon us by our ability to daydream.

15. BUILDING AND REBUILDING THE HUMAN BRAIN

IF YOU WERE TO SWAP YOUR BODY WITH ANOTHER PERSON BUT KEEP THE SAME BRAIN, WHICH ONE OF THOSE BEINGS WOULD THEN BE "YOU"?

Your instinctive answer may be to say that wherever your brain goes, you go: this modest-sized organ is the place where personality is located, where all our past experiences are stored, where emotions and perceptions are registered; in fact, most would agree, it is the very seat of consciousness. What then, if something happened to change that brain – a head injury or a tumour, for example, or maybe dementia – would you then still be *you*? What about if we temporarily change the chemistry of the brain – for example, by taking recreational drugs? Is an altered state of consciousness (*see* Chapter 14) a different "you"? If so, how do we relate this to people who may have temporary episodes of psychosis or depression? We instinctively feel that this 1.3-kg (3-lb) mass of cells inside our head is what makes us who we are, but we also know that it is constantly developing and evolving (*see* especially Chapters 6, 10 and 11), and that it can be changed by trauma or disease. Is it possible and indeed ethical to intervene in this process – for example in the case of autism? To what extent are human behaviours "hard-wired"? Should we use science to rebuild the brain after it is damaged? Will we ever be able to build a machine that is capable of consciousness and empathy? And would we even want to?

Right: We have come a long way since the early days of computers, but will we ever be able to replicate the human mind?

Opposite: Humanoid robots have been developed that can respond to people's feelings but are not themselves capable of feeling.

DESIGNING AND REDESIGNING BRAINS

These philosophical questions have no straightforward answers, but they do provide an interesting perspective on some of the topics explored in this book. For example, although our focus has been on the brain, we have also learnt that the body plays a key role in how we interact with the world (Chapter 13), experience emotions (Chapter 7) and succumb to stress (Chapter 6). Contrary to the beliefs of the seventeenth-century French philosopher René Descartes, the mind and body are intimately connected, and together provide a sense of oneself in the "here and now". In fact, recent research by Yale University's John Bargh suggests that our interaction with the world directly influences our thinking – a phenomenon described as *embodied cognition*. He found that participants judged others as being more caring after they had held a warm drink as opposed to a cold one; that people were more likely to compromise in negotiations if they were sitting in a soft, comfortable chair rather than a hard one; and that individuals took a job more seriously if they were holding a heavy rather than light clipboard. Our body really is part of the way we think and any attempt at artificial consciousness would arguably need to take this into account, as well as other inherently human attributes, such as empathy, emotional sensitivity and creativity.

Setting aside the challenging question of whether it is possible to design a machine that could have a sense of

180 BUILDING AND REBUILDING THE HUMAN BRAIN

self, let us turn to how these philosophical debates inform the potential we have to design or even *re*-design our own brains. Throughout this book we have come across many examples of how an atypical or damaged brain can give rise to particular behaviours or specific difficulties. As our understanding of the biology grows, so do the opportunities to intervene – to "repair" a "broken" brain. However, if we subscribe to the notions outlined at the top of this chapter – that the brain is what gives rise to the self – then it is important to recognize that any physical

Below: Would we want neuroscience to lead the way to some kind of Utopia, where brains are free from disease and there is perfect social cohesion?

Right: When we mirror each other's behaviour, we are showing signs of empathy – an important human trait.

changes to the brain will essentially alter something about the person, which has ethical implications. Simple biological interventions may encompass anything from administering drugs that could influence synaptogenesis, myelination or cell death (*see* Chapter 10), through to psychoactive medications that affect neuronal communication (*see* Chapter 3). Or sometimes hormones can be used as a direct intervention: dispensing the "cuddle hormone" oxytocin to individuals with autism in the form of a nasal spray seems to improve their sociability.

More drastic approaches involve psychosurgery, in which specific regions of the brain are removed (*see* CASE STUDY: HOWARD DULLY, page 187), or may even possibly extend to genetic engineering – the identification and elimination of "faulty" genes. At the extreme end of the equation, one could argue that a biological technique such as this has the potential to produce a brain that is free from disease, with enhanced intellectual capacity, and no desire for war or conflict. A Utopian ideal like this may sound appealing, but it carries many dangers: at what point does a particular behaviour or trait become harmful enough to deserve extinction?; what would the consequences be and how reliably can it be pinned down to a specific genetic code? We would also run the risk of removing traits and characteristics that may play an important role in individual, societal and cultural success. As it happens, some people have argued that conditions like autism and schizophrenia have persisted for a reason – that they have

Below: Some people believe that creative traits may be more common in those who are vulnerable to mental illness.

survived the evolutionary journey only because they are accompanied by traits that are helpful and useful for society, at least in their milder form. Nevertheless, genetic research is hugely important and there is a great deal to be learned by studying inheritance, not least the fact that it can help to clarify the underlying mechanisms of a range of challenging conditions, from Alzheimer's disease to alcoholism.

BUILDING AND REBUILDING OUR OWN BRAINS

Over the last few years, there have been many psychology-related headlines that feature the word *gene*, with suggestions that our DNA can be blamed for, among other things, alcohol consumption habits, infidelity, sexual orientation, delayed puberty, how easily we wake up in the morning, and even political beliefs. From a more medical perspective, there is fairly powerful evidence that genetics contribute significantly to a whole host of conditions, including Alzheimer's disease, schizophrenia, Tourette's syndrome, attention deficit disorder and depression, to name just a few. There has even been fascinating data in the last few years to suggest that psychological *experience* can be passed down the generations: research from New York's Mount Sinai hospital found that the trauma witnessed by people who were exposed to Nazi atrocities during the Holocaust could affect their children's DNA. This is a new and controversial idea, but it ties in well with another recent study in which mice showed inheritance of traumatic memories; male rodents who had been trained to fear the smell of cherry blossom then fathered offspring, who themselves appeared to be immediately afraid of the same smell. This takes the idea of psychological inheritance to a whole new level.

So, let us take a closer look at how nature and nurture come together to produce a unique individual. We know that experience is critically important, but to what extent are we "hard-wired"? For example, why do some people seem to show particular resilience to adversity while others in a supportive environment develop depression? This is probably one of the oldest lines of enquiry in psychology – the extent to which any given human trait or condition is the result of *nature* (inherited and biologically determined) or *nurture* (a result of family, society and

Below: Genes can influence the structure of the brain by changing the instructions for how it is built.

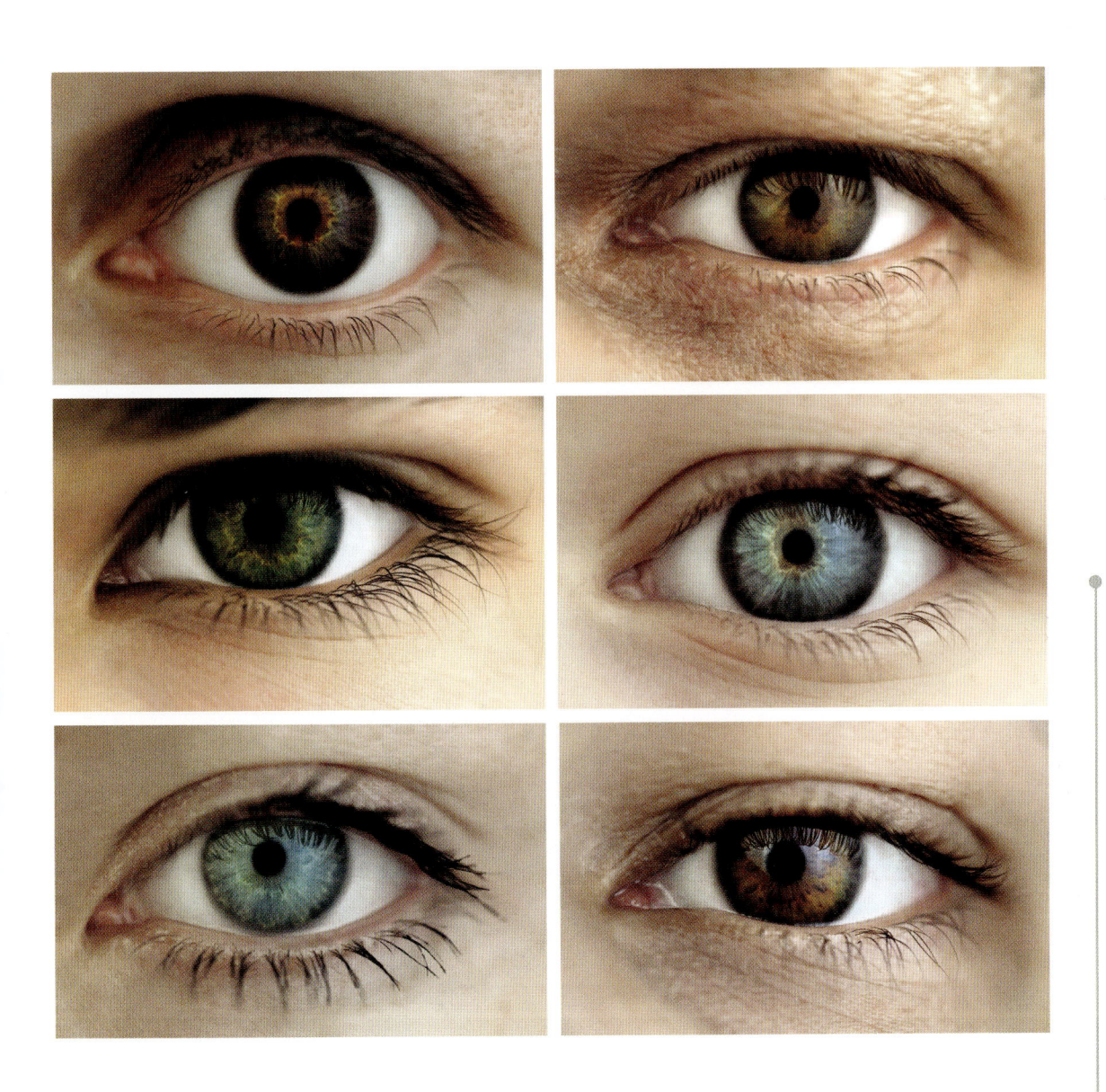

Above: The phenotype refers to the detectable expression of a genotype, such as eye colour.

other environmental influences). Our genetic makeup is stored on strands of DNA inside each of our cells, and this in turn is the product of a specific and unique combination of our parents' own DNA. These genes essentially provide the instructions for protein synthesis. This may sound like an overly simplistic mechanism, but there are around 20–25,000 of these, and amazingly this provides enough flexibility to explain the enormous variability between different human beings.

So, can we isolate the specific genetic codes for conditions such as schizophrenia or personality traits like extraversion? One of the real problems here is that a mental state is difficult to observe. In genetic research we describe relevant observable characteristics – let us say a physical attribute such as eye colour – as the phenotype. This is the detectable expression of a particular genetic sequence. However, when it comes to the expression of a given behaviour, personality or mental functioning, the picture is less straightforward. Some neurological disorders, such as Parkinson's disease, have a relatively reliable diagnosis and clear physical characteristics, making them a good candidate for genetic research. However, others have a far less concrete definition – for example, schizophrenia, anxiety or depression – and concepts such as intellect, mood and political preference are even more slippery! So, despite the enormous recent progress in mapping the entire human genome, when it comes to the genetic basis of behaviour we are still very much in the early stages.

DEFINING GENETIC TERMS

DNA (Deoxyribonucleic Acid): a large molecule that stores genetic code for the synthesis of proteins.
Chromosome: a threadlike strand of DNA bonded to protein; found in nucleus.
Genes: segments of DNA located on chromosomes.
Genotype: the genetic make up of an individual.
Phenotype: the detectable expression of a genotype.

SCHIZOPHRENIA AND GENETIC RESEARCH

Schizophrenia, a distressing psychiatric condition that affects around one per cent of the population, is one mental health condition that has been given particular attention. While diagnosis is not always clear-cut, there is a well-defined set of criteria for the disease and it occurs frequently enough for genetic studies to be possible. In 1996, American psychiatrist Kenneth Kendler led a huge investigation in Roscommon, Ireland. He tracked down all those locals who had been diagnosed with schizophrenia since 1930 and then systematically interviewed all of those people, as well as all their first-degree relatives (a total of more than 2,700 individuals!). He found that people in the immediate family were thirteen times more likely to develop schizophrenia than those without any family connection. While this is interesting, it does not separate nature from nurture – but his findings have since been backed up by evidence from a large number of studies of twins and adoption. The first of these methods compares identical twins (who share the same DNA) with non-identical twins and other siblings; the second examines children who are adopted versus those that stay with their natural family. Both of these approaches have their flaws, but we can afford to have some confidence when they yield similar and consistent results, which they do in the case of schizophrenia. A similar approach has been taken to estimate the genetic contribution to other psychological traits, such as depression (35–40 per cent), personality (35–40 per cent) and alcoholism (50 per cent).

There are two principal messages that we can derive from genetic research. The first is that where inheritance does play a particularly prominent role, we should be looking closely at what the genes might be coding for – for example, are they influencing the structure of the brain? Or the way in which it develops? Or do they affect neural communication directly by altering neurotransmitters or receptors? Secondly, we should be aware that there is always *some* effect of environment, a term that encompasses many factors – from nutrition, exposure to toxins, prenatal stress and birth complications, through to aspects of psychological history, including trauma, education and the level of family support received. Genes carry a set of instructions that are programmed to and only under particular conditions. The challenge is to understand how specific gene-environment combinations influence the way in which the brain is constructed and functions. And although genetic or biological interventions may sometimes be elusive, environment is generally easier to address. For example, we know that Alzheimer's disease progresses more slowly in people who exercise regularly, have low levels of stress and a good diet – and most of these are things that are not difficult to put in place (*see* Chapter 11 for more examples on positive ageing).

Above: Genetic researchers compare the psychological characteristics of identical and non-identical twins.

Finally, what happens when a brain follows a typical, healthy developmental trajectory but then becomes physically damaged due to a traumatic brain injury, stroke, tumour, infection, or poisoning? To go back to the questions at the start of this chapter (*see* page 178), what does this do to a person's sense of who they are? Because we grow very few new neurons after birth, any loss of cells can be quite catastrophic. Some people will simply not survive such an event, while a few lucky ones may escape virtually unscathed. However, it will probably not surprise you to learn that the vast majority of those who *do* survive experience significant changes to the way they understand, process and remember the world. How can science help? Although we do not have the capability to repair brain cells (yet), there are a number of current studies that aim to enhance plasticity, the brain's natural tendency to develop and adapt. These are still in the early stages of development, but such an approach may hold some promise. And for some people, psychoactive stimulants can be useful for sharpening intellect and increasing thought processing speed (*see* page 175). However, the most helpful strategy is to engage in cognitive rehabilitation – an

Above: A wearable camera that takes automatic photographs can stimulate and enhance memory.

Right: Barbara Wilson, founder of the Oliver Zangwell Centre for Neuropsychological Rehabilitation.

approach that combines emotional support for the patient with education and practical strategies, such as the use of a wearable camera.

For now, there is no feasible way either to build or rebuild a human brain and, in reality, the enormous complexity of this precious organ may mean that we never have enough knowledge to make this possible. Nevertheless, we have made massive strides in our understanding since the early days of brain science and there are new and exciting findings occuring every single day. Innovative technology is drawing the focus away from the function of individual brain regions and toward a better understanding of patterns of activation and connectivity. This has the potential to provide a whole new perspective on how the brain gives rise to all our actions, thoughts and feelings. Perhaps President Barack Obama's huge BRAIN project (*see* Introduction) will open up many more doors – or maybe it will leave us with even more questions. In the meantime, I hope that this book has answered some of your own questions and provided fresh insight into what it is that makes you "you". And, most of all, I hope that it has given you a new appreciation of your very wonderful and unique brain.

CASE STUDY: HOWARD DULLY

At 1.30 p.m. on 16 December 1960, Howard Dully became one of the youngest patients to receive a *frontal lobotomy* at the age of just twelve. The procedure was performed by American neurosurgeon Walter Freeman, who had become famous for this pioneering treatment of severe mental health conditions. Freeman's method, based on one that was originally developed by the Portuguese neurologist António Egas Monez in 1936, involved hammering a small kitchen icepick through the tear ducts into the brain and then moving them backwards and forwards to destroy parts of the frontal lobes. Freeman had diagnosed Howard Dully with schizophrenia after his stepmother Lou had complained that he was out of control. In fact, a number of other psychiatrists had previously told Lou that there was nothing wrong with Howard – he was sometimes withdrawn and prone to stealing the occasional biscuit, but by all accounts he was a normal, energetic boy who enjoyed riding his bike and playing practical jokes. In any case, there were a number of mitigating circumstances – Howard's natural mother had died of cancer in 1954, and when Lou came into his stepmother's life she took a quick dislike to him and punished him severely and frequently. After hearing about the possibility of a lobotomy from Freeman, she persuaded him that Howard needed this treatment. However, after the surgery Howard's life went from bad to worse. He moved from one institution to another and eventually ended up a homeless alcoholic involved in petty crime. Amazingly, he eventually managed to get his life back on track: he went to college and got a degree, after which he found a job as a bus driver and then settled down with a wife. He has no disabilities, no obvious emotional difficulties and has above average intelligence. Howard had a lucky escape – although some people with extreme mental health disorders did benefit from the treatment, many were left in withdrawn, zombie-like conditions, while others succumbed to a persistent vegetative state and some even died. Howard had the advantage of the neural plasticity that comes with youth but he still remembers his ordeal, which he describes in detail in his memoirs, *My Lobotomy*, co-authored with Charles Fleming. Howard believes that Freeman had good intentions but was misguided and overly enthusiastic about his procedure. Some psychosurgery is still used today but is carried out with far more precision and caution, and only on those patients for which it is an absolute last resort.

GLOSSARY

Acetylcholine – the first neurotransmitter to be discovered; involved in movement and memory

Action potential – a nerve impulse that travels in one direction, from the cell body to the terminal buttons

Amygdala – an almond-shaped structure found deep in the brain that plays a key role in emotion

Apoptosis – the natural "pruning" away of brain cells and connections that are not being used

Autonomic Nervous System (ANS) – the branch of the nervous system that controls many of our automatic body functions, including heart rate, digestion, pupil size, salivation

Axon – a long thin section of the neuron, which essentially acts as one-way street, carrying nerve signals; bundles of axons are "nerves"

Basal ganglia – a collection of structures deep in the brain that have many connections and are important in movement, emotion and time perception

Broca's area – a region in the frontal lobes of the brain that is important for the production of speech; first identified by Paul Broca

Dopamine – a neurotransmitter that plays a key role in our sense of reward and pleasure; also important for movement, learning and memory

Central Nervous System (CNS) – comprises the brain and the spinal cord; processes all incoming sensory information and coordinates conscious and unconscious behaviour

Cerebellum – a large cauliflower-shaped structure that hangs off the back of the brain; important for movement, balance, emotion, attention and time perception

Cerebral cortex – the outer layer of the brain, which is responsible for thinking; divided into four lobes – the frontal lobe, parietal lobe, temporal lobe and occipital lobe

Computerised Tomography (CT) – a technique that uses a computer to collate data from many X-rays in order to provide a picture of the brain and other internal organs

Cortisol – a hormone that enables our body to respond to stress/change; sometimes referred to as a "stress hormone"

Electroencephalogram (EEG) – a reading of electrical activity in the cerebral cortex; gives an indication of alertness

fMRI – a functional imaging technique that uses strong magnets to assess blood flow in the brain, which in turn gives an indication of which parts are active

GABA – a neurotransmitter that helps to inhibit (dampen) neural activity; its activity is enhanced by alcohol and tranquilizers

Glial cell – brain cells that support and protect the neurons

Hallucination – an experience of seeing, hearing or smelling something that isn't present in the environment

Hippocampus – a sea-horse shaped structure found towards the middle of the brain; plays a key role in memory

Hypothalamic Pituitary Axis (HPA) – the "slow" stress pathway; coordinates the body's hormonal response to stress

Hypothalamus – a small region of the brain that controls the autonomic nervous system and pituitary gland; plays a vital role in keeping internal body conditions constant

Limbic system – a set of central brain structures that plays a key role in memory and emotion

Magnetic Resonance Imaging (MRI) – a technique that allows us to see the brain and other internal structures in great detail

Myelination – a developmental process where insulation is added to neurons enabling them to carry nerve impulses faster and further

Neurogenesis – the birth of new neurons

Neuron – a specialized cell that is capable of carrying electrochemical signals around the body and brain

Neurotransmitter – a chemical messenger that allows neurons to communicate with each other

Noradrenalin – a neurotransmitter that plays a key role in the emergency "fight-or-flight'" response and affects our overall arousal

Oxytocin – a hormone that is important for bonding, attachment and love; sometimes called the "cuddle hormone"

Peripheral Nervous System (PNS) – the branch of our nervous system that carries sensory information in and movement instructions out of the brain

Pituitary gland – a small pea-sized structure at the base of the brain, known as the "master gland" of the endocrine (hormone) system

Plasticity (or "neuroplasticity") – the physical shaping of our brain in response to our environment and experience

Positron Emission Tomography (PET) – a technique that enables us to assess brain activity by measuring uptake of radioactively labeled glucose

Prefrontal cortex – the outer layer of the frontal lobe of the brain; important for memory, planning, behavioural control, empathy, personality and consciousness

Serotonin – a neurotransmitter that is important for many body functions, including eating, temperature regulation, sleep and mood

Suprachiasmatic nucleus (SCN) – a region of the brain that uses light to keep our brain and body in sync with the day-night cycle

Sympathomedullary pathway (SAM) – the "fast" stress pathway; activates the "fight-or-flight" response in times of acute stress

Synaesthesia – a condition in which the senses become mixed; for example, sounds may also be seen as colours or shapes, or tastes may have a sound

Synapse – the junction between two neurons

Synaptogenesis – the development of new connections between neurons

Wernicke's area – a region of the brain that is important for the understanding of speech

INDEX

(page numbers in italic type refer to illustrations)

acetylcholine 33, *33*, 36, 39 (*see also* neurotransmitters)
    and acetylcholinesterase inhibitors 39
ADHD, *see* attention deficit hyperactivity disorder
adrenaline 54, *55*
agonists 37, *45*
alexithymia 96
Ali, Muhammad *38*
Alzheimer's disease 15, 33, 39, *39*, 66–7, *66*, 84, 123, 134, 141, 182, 185
amnesia, *see under* memory
amphetamine 48
amygdala 72, *73*, 79, 81, 84, 92–5, *95*, *96*
    and music 110, *112*
amyloid plaques 134, *134*
anorexia 82, 96, 123, 159
antagonists 37, *45*
anterior cingulate 92, *105*
antidepressants 40–3, *41*
anxiety disorders 42
apoptosis 123
Aristotle 12, 88, 156
Asperger, Hans 85
Asperger's syndrome 84
attention deficit disorder (ADD) 60, 84, 102, 182
attention deficit hyperactivity disorder (ADHD) 123
auditory cortex 110, *112*, *147*
autism 84–5, 91, 123, 181–2

Bard, Philip 91
Bargh, John *179*
Baron-Cohen, Simon 84–5
basal ganglia *21*, 23, 92, 104, *105*, 106, 107
basal nucleus 39
Bell, Alexander Graham 165
Bell, Vaughan 33
benzodiazepines 42–3, 48 (*see also* antidepressants)
Berger, Hans 13
beta-blockers 40, 42 (*see also* antidepressants)
Blakemore, Sarah-Jayne 129
Blakemore, Susan 169
brain (*see also* neurons; neurotransmitters; *individual parts and mental conditions*):
    action potential in *28*, 29, *32*
    ageing 132–41, *133*, *134*, *135*,
    136, *137*, *138*, *139*, *140*, *141* (*see also* memory)
    building and rebuilding 182–4
    changing chemistry of 36–8, *37*
    chemical-imbalance theory of 51
    designing and redesigning *178*, *179*–82, *179*, *180*
    examining structure of 15, *15*
    gender/sex differences in 76–87, *76*, *77*, *78*, *79*, *80*, *81*, *82*, *83*, *84*, *85*, *86*
    and genes 123, 181, 182–5, *182*, *183*, *185*
    grey matter 71, 123, 129, *136*
    pathways through 34–5, *54*, *55*, 56
    plasticity of 71, 120, 127
    responses of, to stress 52–6
    and time, *see* time
    training 72
    white matter 123, 124, 129
    young and developing 120–31, *120*, *121*, *122*, *123*, *124*, *125*, *126*, *127*, *128*, *129*, *130*, *131*
Brain Gym 131
BRAIN Initiative 8, 187
Broca, Paul 16–17, 92, 145, 146
Broca's aphasia 17, 146, *147*
Broca's area 17, 146, *147*, 149
Buñuel, Luis 65

caffeine 48, 63, 175
cannabis 49, *51*
Cannon, Walter 91
caudate nucleus 110, *111*
cell membrane 26, *26*
central nervous system, *see* nervous system
cerebellum 21–3, *21*, 67, 72, *73*, 92, 104, *105*, 106
    and music 110–13, *112*
cerebral cortex *21*, 23, 69, *73*, 84
cerebrum 23
Chabris, Christopher 75
chlorpromazine 40, 42 (*see also* antidepressants)
cingulate gyrus *94*
circadian rhythms 48, 63, 102–3, *103*
Clow, Angela 59
cocaine 48
Cognitive Behavioural Therapy (CBT) 42
cognitive rehabilitation 186–7
Computerized Axial Tomography (CT/CAT) 15
Computerized Tomography (CT) 13, 15–16, *16*
Confucius 108
consciousness, altered states of (ASCs) 166–77, *166*, *167*, *168*, *170*, *171*, *172*, *174*, *175*, *176*, *177*
    and arousal, attention and memory 169
    deliberately induced 173–6, *174*
    hypnagogia and hypnopompia 171–3
    and hypnosis 167
    out-of-body and near-death *172*, 173
    and sleep and dreaming 170
    spontaneous and physically induced 170
Conway, Martin 74
cortisol 54, 56, 58–9, *58*, *59*, 60, 61 (*see also* hormones)

Dale, Sir Henry 33
Damasio, Antonio 89, 92
Damasio, Hanna 89, 92
Darwin, Charles 91
Davies, John Booth 116
Dennison, Gail 131
Dennison, Paul 131
depression 20, *20*, 23, 34, 39, *42*, 51, 58, 65, 96, 102, 123
Descartes, René *179*
Deter, August *39*
DNA, *see* brain: and genes
Dodsen, John 55
dopamine 33–4, 40, 42, 48, 109 (*see also* neurotransmitters)
    and Parkinson's 38–9
Droit-Volet, Sylvie 100
drugs 46
    and addiction, dependence, tolerance, withdrawal 43–5, *43*
    amphetamine 48
    and brain receptors *44*, 45
    caffeine 48, 63, 175
    cannabis 49, *51*, 174
    cocaine 48
    ecstasy 48 (*see also* drugs: recreational)
    effects of, on brain 36–51, *37*, *41*, *42*, *44*, *45*, *46*, *50*
    hallucinogens 49, *50*, 174–6
    and illegal highs 48–51
    LSD 49, *50*, 174, 175
    magic mushrooms 49, *102*, *177*
    in mental health 39–43, *42* (*see also* antidepressants)
    mescaline 176
    morphine *49*
    nicotine *43*, 48
    psychedelic *50*, 51, 174–6, *175*
    recreational *43*, 47–51, *48*, *50*, 174–5
    and reward pathway *46*
Dully, Howard (case study) 187
dyslexia 123

Eagleman, David 100, *101*, 102
Eccles, Sir John 25
ecstasy (drug) 48 (*see also* drugs: recreational)
Edginton, Trudi 97
Einstein, Albert 98 (*see also* time)
electroencephalogram (EEG) 13, 18–19, *19*, 173
    and music 110
emotions 88–97, *88*, *89*, *90*, *91*, *92*, *93*, *94*, *95*, *96*, *97*
    Cannon–Bard theory of 91
    and music *112*, 116–18, *116*, *117*, *118*
    theories of *93*
endocrine system 56–7, *57* (*see also* hormones)
endorphins 109
Evans, Vyv 89

fight-or-flight response 12, 42, 48, 55, 91 (*see also* emotions)
Fine, Cordelia 83, 84
Fleming, Charles 187
Forrester, Gillian 143
Fox, Michael J. 38–9, *38*
Frassinetti, Francesca 101
frontal lobes 18, *22*, 23, 35, *73*
    and ageing *133*, 135

GABA 34, 42, 48 (*see also* neurotransmitters)
Galen 12
Gall, Franz Joseph 13–14, *13*
Geschwind, Norman 79, 85
glial cells *24*, *24*, 120
Glover, Vivette 60
Goldacre, Ben 131
Gray, John 77, 84
Greenough, William T. 127, 128
Gurney, Daniel 151–2

hallucinogens 49, *50*
Halpern, Diane 84
Hammond, Claudia 106
Hebb, Donald 69
Hendrix, Jimi *43*, 156
heroin 48–9
hippocampus 61, 67, 72, *73*, 79, 81, 92
    and ageing 135

INDEX

HM (Henry Molaison) 65–6, 67, 72
Hoagland, Hudson 101
Hodgkin, Alan 25, 26, 29
Hofmann, Albert 175
Holocaust 182
hormones 55–8
    prenatal 85–7, *85*, *86*
    sex 84, 130 (*see also* brain: and gender/sex)
    "stress hormone", *see* cortisol
Hornykiewicz, Oleh 38, 39
Hubel, David 126–7, *126*, 128, 159
Human Genome Project 8
Huxley, Aldous 176, *176*
Huxley, Andrew 25, 26, 29
hypothalamic-pituitary-adrenal (HPA) system 55
hypothalamus 21, *21*, 54, 55, 58, 92, *94*, 95

insula 92
insulin 58 (*see also* hormones)
ions 26–9, *26*, *27*

James, William 91, 120, 165
Jansari, Ashok 97
Jarrett, Christian 84
Joel, Daphna 81
Juslin, Patrick 108

Keller, Helen 165
Kendler, Kenneth 184
Konorski, Jerzy 120

L-dopa 39, 43 (*see also* neurotransmitters)
Lange, Carl 91
language 17, *127*, 142–53, *142*, *143*, *144*, *145*, *146*, *147*, *148*, *149*, *150*, *151*, *152*
    and aphasia 17, 145–9, *146*, *147*
    Braille *145*, 158
    developing 143–4
    development of, in children 150, *151*
    non-verbal communication 149–52, *150*
    prenatal acquisition of *142*
Lashley, Karl 69
Laukka, Petri 108
LeDoux, Joseph 90, 92
Leonardo da Vinci 10
Levy, Becca 141
limbic system 23, 92, *94*
    and music 110, *112*
Lindquist, Kristen 95
locus coeruleus *133*
Loewi, Otto 33
LSD 49, *50*

MacDonald, John 151

McGurk, Henry 151
McPhie, Leland 132
Magnetic Resonance Imaging (MRI) 13, 15, 16, *16*, *17*, 129, *136*
    functional (fMRI) 18–19, *20*
Maguire, Eleanor 71
Manning, John 87
Meck, Warren 104, 107
meditation 173–5
memory 64–75, *65*, *66*, *68*, *69*, *70*, *71*, *73*, *74*
    and ageing brain 135–7 (*see also* brain: ageing)
    and altered states of consciousness 169
    anatomy of 72–4
    brain parts relevant to *73*
    and building connections 69–71
    declarative and non-declarative 67
    episodic 64
    everyday 74–5
    long- and short-term 67
    loss of 65–7
    "muscle" 72, *74*
    non-unitary theories of 67
    storage of *68*, 69
    and time 106–7, *107* (*see also* time)
Milner, Dr Brenda 66
Moffat, Steven 98 (*see also* time)
morphine 48–9
Morris, Chris 102
motor cortex *147*
Moulin, Chris 74
music 108–19, *108*, *109*, *111*, *112*, *113*, *114*, *115*, *116*, *117*, *118*, *119*
    and animals 113, *113*
    in brain 110–13, *111*, *112*
    chill response to 109, 110, *111*, 113
    and cultural identity 109, *114*
    and emotions *112*, 116–18, *116*, *117*, *118*
myelination 120, 123, 181

Nash, John *84*
nervous system 30, 46, *56*
    autonomic (ANS) *11*, 12, 55, 61, 90, 92, *93*
    organization of 11–12, *11*
    peripheral (PNS) 12, *12*
neurofibrillary tangles 134, *134*
neurogenesis 120–3, *121*
neurons 24–9, *25*, *31*
    neurotransmitters' effects on 37
neuroplasticity 71, 120, 127
neuropsychology 17–18
neurotoxins 36
neurotransmitters 30–5, *31*, *32*,

33, *35*, *37* (*see also* acetylcholine; dopamine; GABA; noradrenaline; serotonin)
    and antidepressants 40
    and synapses 37
nicotine *43*, 48
noradrenaline 33, 34, 40, 42, 48 (*see also* neurotransmitters)
North, Adrian 113
nucleus accumbens 47, 110, *111*

Obama, Barack 8
obsessive compulsive disorder (OCD) 23, 42
occipital lobes *22*, 23
oestrogen 56, 85, 87, 130
oxytocin 109, 159

Panksepp, Jaak 109, 113, 114–15, 118
Papez, James 92
parahippocampal gyrus 110, *112*
parietal lobes *22*, 23
Parkinson's disease 14, 38–9, *38*, 65, 102, 104, 123
Parsons, Lawrence 110
Pascalis, Olivier 125
Penfield, Wilder 13, 47
phrenology 13–14, *14*
Pinker, Stephen 118, 144
pituitary gland 21, *54*, 55–6, 55, 58
Plato 88
Positron Emission Tomography (PET) 13, 18, *149*
    and music 110
Pratchett, Terry 141
prefrontal cortex (PFC) 23, 67, 72, 92, *105*, 106, 129, *168*
    and ageing 135
pruning 120, 123, *123*

reward pathway 47, *47*
Ridge, Damien 40
Rosenzweig, Mark 71, 127
Rutter, Michael 128

Schachter, Stanley 91–2, 95
schizophrenia 14, 39, 42, 84, *84*, 123, 181–2, 184, 187
Scott, Sophie 35, 89
Selective Serotonin Re-uptake Inhibitors (SSRIs) 40, *41* (*see also* antidepressants)
Selye, Hans 52, 62
sensation and perception 154–65, *154*, *155*, *157*, *158*, *159*, *160*, *161*, *162*, *163*, *164*, *165*
sensory cortex 156, *160*
serotonin 33, 34, *34* (*see also* neurotransmitters)
    and SSRIs 40, *41*

Shebalin, Vissarion 110
Siffre, Michel 102–3
Simons, Daniel 75
Singer, Jerome 91–2, 95
sleep 63, 140, *140*, 169, 170
    and dreams 170
Sloboda, John 117
Slocombe, Katie 143
somatosensory cortex 158
Sommer, Iris 79
Standley, Jayne 109
stress 52–63, *52*, *53*, *54*, *55*, *56*, *57*, *58*, *61*, *62*
    "stress hormone", *see* cortisol
striatum 107
stroke 15, 20, *20*, 65, 110, 123
substantia nigra 38
    and ageing *133*
suprachiasmatic nucleus (SCN) 103, *105*
Sympathetic Adrenomedullary (SAM) system, *see* fight-or-flight response
synaesthesia 156
synapses *25*, 30, *31*, *32*
    and memory 69–71, *70*
synaptogenesis 120, *122*, 123, 181

Tart, Charles 167
temporal lobes 18, *22*, 23, 65–6
testosterone 79, 85, *86*, 87, 92, 130
thalamus 21, *21*
time 98–107, *98*, *99*, *100*, *101*, *102*, *103*, *104*, *105*, 107
    and internal clocks 102–6, *105*
    and memory 106–7, *107* (*see also* memory)
    understanding 98–100
Tourette's syndrome 23, 84, 123, 182
Transcranial Magnetic Stimulation (TMS) 20–1, *20*
tricyclic antidepressants (TCAs) 40 (*see also* antidepressants)
Tyler, Lorraine 132

Vaitl, Dieter *166*, 170

Wearing, Clive 65–6, 67, 75, 119, *119*
Wedekind, Claus 162
Wernicke, Karl 146
Wernicke's aphasia 146, *147*
Wernicke's area *147*, 149
Wiesel, Torsten 126–7, *126*, 128, 159
Wilson, Barbara *186*

Yerkes–Dodson Law *53*, 55
Yerkes, Robert 55

Zatorre, Dr Robert 110

CREDITS

The publishers would like to thank the following sources for their kind permission to reproduce the pictures in this book.

Alamy: 115 This Life Pictures

Corbis: 126 Ira Wyman/Sygma

The Encephalitis Society: 186 (bottom)

Getty Images: 9; /114 Timothy Allen: /84, 179 ChinaFotoPress; /180 Chris Clor; /152 Education Images/UIG; /176 Edward Gooch; /165 Viviana Gonzalez; / 38 Harry Hamburg/NY Daily News Archive; /65 Andrej Isakovic/AFP; /129 Cynthia Johnson/Liaison; /88 Andreas Kuehn; /62 (top left) Ronnie Kaufman/Larry Hirshowitz; /Donald Kravitz: 149; /82 Caiaimage/Tom Merton; /91 Cordelia Molloy; /69 Dana Neely; /Massimo Pizzotti: 164; /109 Oli Scarff/AFP; /25 Science Picture Mo; /108 (right) Phil Sills Photography: /178 SSPL; /130 Tetra Images; /174 Tannis Toohey/Toronto Star; /62 (centre left) Betsie Van Der Meer; /180 (top) Jim Watson/AFP

iStockphoto.com: 99, 107, 123, 184-185

Microsoft Research Cambridge: 186 (top)

Ben Mills: 58

NASA: 171

Public Domain: 155

REX Shutterstock: 119 John Dee; / 117 Alastair Muir

Science Photo Library: 10, 35, 39, 48, 137; /15 AJ Photo; /167 Abk/BSIP; /19 (top), 20 (bottom), 120 Amelie-Benoist; /102 Martin Bond; /162 Henning Dalhoff; /46 Fernando Da Cunha/BSIP; /144 Nick David; /146 Thomas Fredberg; /83 Ian Hooton; /1, 24 Hybrid Medical Animation; /131 Laguna Design; /158 Claus Lunau; /104 Living Art Enterprises; /136 Dr P. Marazzi; /135 Medical RF.Com; /12 Mikkel Juul Jensen; /30 Mehau Kulyk; /13 National Library of Medicine; /16 (right) Medical Body Scans; /160 Medical Images, Universal Images Group; /16 (left) Miriam Maslo/SPL; /33 Alfred Pasieka; /18 Lea Paterson; /142 Redheadpictures/Cultura; /134 Martin M. Rotker; /68 Sciepro ; /19 (bottom), 80 Sovereign, ISM; /55 Bob L. Shepherd; /51 Adrian Thomas; /148 Wellcome Dept. of Cognitive Neurology; /14 Ken Welsh/Design Pictures; /17, 20 (top) Zephyr

Shutterstock.com: 2, 34, 42, 43, 47, 49, 50, 52, 56, 59, 60, 61, 66, 71, 74-75, 76-79, 81, 85, 86, 89, 90, 92, 95, 97, 98, 100, 101, 103, 104 (left), 113, 116, 118, 124, 127, 128, 138, 139, 141, 143, 145, 150, 151, 154, 159, 161, 162 (top), 166, 170, 175, 177, 181, 182, 183

Illustrations by Phil O'Farrell.

Every effort has been made to acknowledge correctly and contact the source and/or copyright holder of each picture and Carlton Books Limited apologises for any unintentional errors or omissions, which will be corrected in future editions of this book.